Praise for *Cradle to Cradle*

"One of the most influential recent books on design and environmentalism." —Alice Rawsthorn, *The New York Times*

"[McDonough] point[s] to a path out of the seemingly unwinnable trench war between conservation and commerce." —James Surowiecki, *The New Yorker*

"A rare example of the 'inspirational' book that actually is." —Steven Poole, *The Guardian*

"[McDonough and Braungart's] ideas are bold, imaginative, and deserving of serious attention." —Ben Ehrenreich, *Mother Jones*

"[A] clear, accessible manifesto . . . The authors' original concepts are an inspiring reminder that humans are capable of much more elegant environmental solutions than the ones we've settled for in the last half-century." —*Publishers Weekly*

"A readable, provocative treatise that 'gets outside the box' in a huge way. Timely and inspiring." —*Kirkus Reviews*

"For those of us who have a hunger to know what the next great idea will be, this highly readable book captures and challenges the imagination." —Sarah D. Scalet, *OnEarth*

D0029476

Also by William McDonough and Michael Braungart

Cradle to Cradle: Remaking the Way We Make Things

The
Upcycle

The
Upcycle

Beyond Sustainability—
Designing for Abundance

William
McDonough
and
Michael
Braungart

MELCHER
MEDIA

NORTH POINT PRESS
A division of Farrar, Straus and Giroux
New York

North Point Press
A division of Farrar, Straus and Giroux
18 West 18th Street, New York, NY 10011

Printed in the United States of America
First edition, 2013

Text and Image Credits:
p. 47: reprinted from *A World on the Wane* by Claude Lévi-Strauss, trans. John
Russell (London: Hutchinson, 1961). Copyright © 1955 Editions Plon; p. 57:
© Corbis; p. 126: courtesy of Conservation Research Institute/Heidi Natura

Library of Congress Control Number: 2013933161
ISBN: 978-0-86547-748-3

Produced by Melcher Media
124 West 13th Street
New York, NY 10011
www.melcher.com

Publisher: Charles Melcher
Associate Publisher: Bonnie Eldon
Editor in Chief: Duncan Bock
Editor: Elizabeth Mitchell
Production Director: Kurt Andrews
Digital Production Associate: Shannon Fanuko
Editorial Assistant: Lynne Ciccaglione

Designed by Office of Paul Sahre

www.fsgbooks.com • www.twitter.com/fsgbooks • www.facebook.com/fsgbooks

10 9 8 7 6 5 4 3 2 1

To our families,
and to all of the children,
of all species, for all time

Glance at the sun.
See the moon and the stars.
Gaze at the beauty of earth's greenings.
Now,
Think.
—Hildegard of Bingen

Betrachte die Sonne.
Sieh den Mond und die Sterne.
Erkenne die Schönheit der Natur.
Und dann denke nach.
—Hildegard von Bingen

If I were given one hour to save the planet, I would spend
59 minutes defining the problem and one minute resolving it.
—Albert Einstein

We have this remarkable experience in this field of fundamental
physics that beauty is a very successful criterion for choosing the
right theory. Why on earth could that be so?
—Murray Gell-Mann

The goal of the upcycle is a delightfully diverse, safe, healthy, and just world with clean air, water, soil, and power—economically, equitably, ecologically, and elegantly enjoyed.

Contents

Foreword

I first met Bill McDonough in the early 1990s, when he brought some remarkable design ideas to Greening the White House, an initiative I launched to dramatically reduce the White House's energy consumption and make it a model of efficiency. Bill, an American architect, had just teamed up with the German chemist Dr. Michael Braungart to write the Hannover Principles, which were already becoming an international touchstone in green circles. This set of ideas, about how to design safe cities, homes, and workplaces, and how to endlessly reuse the earth's resources more efficiently and more effectively, struck me as something bigger than an academic exercise. These ideas made sense, and they were doable.

Bill and Michael proposed that a better-designed world would be good for business, good for people's health, and good for the environment. Their first book, *Cradle to Cradle: Remaking the Way We Make Things*, introduced these ideas to the broader public and gave momentum to the sustainability movement, urging us to eliminate the concept of waste and arguing that no resource ought to be considered dispensable. I've watched as many of the concepts presented in *Cradle to Cradle* have taken root at the U.S. Postal Service and NASA, at small businesses and corporations as large as Walmart and Procter & Gamble, and in countries all over the world. I've seen how these simple ideas, when put into practice, can improve productivity and make people happier and healthier.

In 2008, I visited Make It Right, the program Brad Pitt founded with Bill's consultation to help Hurricane Katrina victims return home to New Orleans's devastated Lower Ninth Ward. The program's designers and builders were applying Cradle to Cradle principles and processes throughout the construction of the new houses. A few years later, I heard from a woman who had spent three years in emergency housing in Texas but had finally returned to New Orleans thanks to Make It Right. She had a daughter who had always wanted to take dance lessons. After she'd moved into her new, healthier, low-cost home, not only did lower utility bills enable her to afford some lessons, but her daughter's once-severe asthma disappeared because of the cleaner interior materials. She could breathe again—and dance.

That is the essence of Bill and Michael's work—the genuine desire to help others, coupled with intellectual curiosity and a deep commitment to constant improvement. They work to transform "good enough" into the very best. They focus on making the right things the right way.

After a long career in elected office and more than a decade traveling the globe for the Clinton Foundation, I've learned that we get the best outcomes when we make decisions that are rooted in evidence and experience—when we put aside ideology and focus on what works. The ideas that Bill and Michael put forward in this book come from an honest sensibility that transcends the daily finger-pointing of left, right, or even reverse. They just point forward.

The Upcycle is a book about creativity, about thinking big even if we have to act small, and about approaching problems with a bias for action. It encourages us to find solutions through close observation, innovation, and the study of real, local conditions and needs. This is the approach that has made Bill and Michael's work so effective over the years—whether it's working to design a super-efficient building with NASA, partnering with some of the world's biggest companies to devise renewable products and energy systems that are good for the bottom line, or helping victims of Hurricane Katrina get a new start in better, more healthful houses.

The optimist says the glass is half full and the pessimist says it's half empty. Bill and Michael say it's always totally full—of water and air—and they are constantly working to share that full glass with more people, to make it even bigger, and to celebrate the abundance of the things that enable us to thrive.

In the pages that follow, Bill McDonough and Michael Braungart invite you to think about the future we share; to imagine what could be and how to make it so. We are all in this together, and we'll need a global commitment to sustainability if we want our children to inherit a world of shared opportunity, shared responsibility, and shared prosperity. Let's get to work.

—President Bill Clinton

The
Upcycle

Introduction

Imagine you are sitting in the top-floor boardroom of a major United States consumer products company and you are meeting one-on-one with the company's executive in charge of sustainability. You have been to this facility many, many times before. Over seven years, you have met with executives in charge of finance, supply chains, manufacturing, product design, research and development, and marketing. Hundreds of meetings to listen, to learn, and to explore your new concepts for sustainable growth and beneficial innovation.

Together, you and the executive have shared data—lots of data. You know big-picture business issues facing this company and detailed chemistries of the products. You even know how many light-bulbs are used to illuminate the enterprise worldwide, how much energy that consumes, how many lightbulbs contain mercury, and how many people it takes to change a lightbulb and what that costs.

This is the nature of the work. To use a detailed, defined inventory as a platform for invention, innovation. To ask and answer: What's next?

Outside the giant plate-glass windows, tall granite-clad skyscrapers stand proudly in the sunshine. The Brazilian mahogany table is polished to a shine, and the high-backed leather chairs remind you of the important executive decisions made in this room, which can affect the lives of millions of people—for better or for worse. One might say you are here chasing the butterfly effect. Given the scale of this company, one small decision has the power to make a real difference for the economy, for people, and for the planet.

That is one reason you are here—scale. But you are also here for another reason—velocity. Many of the largest corporate enterprises in the world have come to realize the downside of the butterfly effect, the repercussions of modern business that are obviously damaging and too often unaccounted for—famously called externalities, such as carbon in the atmosphere, toxic materials, poisoned rivers, lost rain forests, and so on, with no end of this decline in sight. Many businesspeople realize this is not good business. They like to know what they are doing and to be able to account for it, but they feel like they are driving a car without a gas gauge or even, shall we say, a battery charge indicator? It makes them nervous.

They also are like Olympic athletes who want to be on a safe, level playing field and who do not want to be left behind. They want to lead.

You might just ask this executive friend, "Wouldn't it be wonderful if you could commit not just to reducing your carbon emissions but to being 100 percent renewably powered? Couldn't we find a way to make such a statement?"

The executive pushes the question aside. "We can't do that," he says. "No matter how much we would like to declare ourselves that way. Look, we could only get a small percentage of our power for our factories from solar on our roofs. We and everyone else have been saying we'll cut our carbon emissions 20 percent by 2020. Isn't that enough? Because of the nature of business, we have to be conservative and risk averse. We can only describe actual performance goals that are realistic. How in the world can we say we are going to seek renewable energy for our entire global enterprise? Consumers don't care and environmentalists won't trust us, or if we launch the initiative piecemeal—which is the only way we could—the public awareness of the issue will become a point of concern for all the other products made by the company. For example, if we say these plants are renewably powered, it will raise the question of 'Why not the other ones?' and it's a big, long job getting there. Our shareholders will think we've lost sight of our revenue and profit goals."

"What if you just state your intention?" you suggest. "Say, 'We will be renewably powered as soon as it is cost-effective, and we will constantly seek it out.' Any shareholder can understand that plan. It's true, and declaring your intention does the heavy lifting of getting people in the company to get moving in this direction. You've charted the goal. You'll track your progress and report it. You'll unleash the creativity and genius of your people in a clear, clean direction. You've made them want to search for the renewable power solution every time they go looking to supply a kilowatt-hour. It lets other industries know that if they can manufacture the solar panels or wind turbines or biogas collectors at a competitive price point, they will have a customer in you. And you, their customer, are likely to lead to other major customers. Before you know it, the renewable power industries are growing technologies and jobs in a businesslike way all across the United States, around the world. Your intention itself is powerful."

"Okay, I get it," the executive says. "I'll put this in terms the business will understand and take this to the CEO."

This story actually happened. We didn't have to look far to see how this was just one executive in this mammoth company, against the endless horizon of people in offices outside the plate-glass windows. This was one person, but this person could take a message to the leadership that would launch innovation as inspired as sending a person to the moon. In a few short months, the company announced it would pursue the goal of being renewably powered. All kinds of marvelous innovation busted loose within days. Factory managers started calling, saying, 'Can I go first?' 'What can I do to get on board?' Velocity.

We tell this story without names for two reasons. The first is that this is not a unique story and the point of telling it is to focus on and celebrate the power of intentionality. We know that everyone—consumers, manufacturers, government leaders—is interested in a cleaner, healthier world. Many companies with whom we work are delighted to embark on creating a renewable energy base. They would also love to make their products with only fully defined healthful materials. But society has factionalized to become so mutually suspicious that often consumers and customers don't think companies want the same positive healthful future they want, and companies think critics will pounce on them if they even lift their heads to break out of the norm to say, "We want to try. We are trying. We have embarked on the work of being renewable or pursuing only clean production or fully healthful products, but we have more work to do."

We hope this book, if nothing else, will inspire you to start and will cheer you on. We believe in constant improvement. Sometimes you can't do it. It doesn't work. Fine. Try another way—do it again and again. Restate your intention. Watch what happens.

Secondly, we tell this story without names because we want you to see yourself as both of these individuals in the conference room. All it took was one advocate and one executive to craft a strategy and to move it toward the head of the company, and an entire international corporation was changed. That person could be you in your job, your daily life. When we say "start," we also want you to start thinking of yourself as a potential leader. As a person who changed his or her company, home, country for a better, more beneficial future. This book is written for you. We hope to encourage and inspire you with the how and why of creating a more abundant, joyful world for future generations.

Now we want to tell a story that names names. We were in a meeting with Walmart talking about how they could keep improving their environmental footprint in the world. Walmart is intent on being 100 percent renewably powered and has stated so very publicly; they are now the largest corporate users of solar collectors in the United States. Walmart is also working with local food growers to cut down on shipping vegetables long distances, thus making air and atmosphere cleaner. Of course they have a great deal of work to do, but they are starting and well under way. We were talking about products, how to use only positively defined ingredients in packaging in goods sold in the stores. The question on the table was "Is it really possible?" Just then, we noticed a United States Postal Service Priority Mail box on the desk behind the executive. "Let's turn it over," we said.

The executive did just that.

There was the little certification stamp: Cradle to Cradle Certified^{CM}.

That meant that the mailer had gone through our certification process to identify the chemicals and processes used to make the product, and that the product was fully defined—meaning we had identified and assessed every ingredient—and was on its way to its beneficial optimization of materials, logistics, energy, water, and social fairness. There we were observing Walmart observing USPS's commitment to being healthful.

In China, Goodbaby, the world's largest maker of childrens' products, already has published a road map showing how it will adopt Cradle to Cradle standards company-wide. We name these names to show you that the most mainstream companies in the world are thinking in this positive way. We think you can too.

This philosophy, and this book, are the upcycle of our previous work.

A decade ago, we—Bill, an architect, and Michael, a chemist—published *Cradle to Cradle: Remaking the Way We Make Things*. We had come across an idea in our design and chemistry work that we considered extraordinarily exciting. Human beings don't have a pollution problem; they have a design problem. If humans were to devise products, tools, furniture, homes, factories, and cities more intelligently from the start, they wouldn't even need to think in terms of waste, or contamination, or scarcity. Good design would allow for abundance, endless reuse, and pleasure.

This concept, we believe, could move the dialogue far beyond a simple interest in recycling, because we noticed that the entire recycling effort grew from a negative belief. The theory being put forward by most sustainability advocates, and increasingly by industry, goes something like this:

Human beings create enormous amounts of waste and should strive to become "less bad." Use less energy. Poison less. Cut down fewer trees. According to these current "best practices," all people can hope to achieve is eco-efficiency, minimization, and avoidance, to recycle a limited percentage of objects humans use daily—bottles, paper—and fashion them into, unfortunately, a lesser product, one that can be used once more, or twice more, or maybe even five times more. But then where does this product go? Into a landfill? An incinerator?

That might not be so bad if the product were well designed from the first. It could become a nutrient in the biosphere. Or stay in the technosphere—as a reusable metal or plastic—instead of contaminating the biosphere, the entire ecosystem.

This project, as big as it sounds, is obviously not impossible: Nature itself designs this way.

But as modern engineers and designers commonly create a product now, the item is designed only for its first use, not its potential next uses after it breaks, or grows threadbare, or goes out of fashion, or crumbles. The item works its way from one downward cycle to another, becoming less valuable (think a food-grade plastic bottle smashed down, remelted with other plastics, and made into a speed bump) or more toxic (such as wood turned into a composite board made of formaldehyde-based glues). It seems that what humans make is detritus, frequently toxic.

We believe there is a different perspective. The problem is not with humans per se, but with what they have in the last 5,000 years, and especially in the last 150 years, fashioned.

When the Industrial Revolution manifested itself, people wanted simply to keep supply as high as demand, and as they did so, thinking grew frantic. Designers and manufacturers grasped for the next best short-term idea that came along, not necessarily informed by long-term considerations.

Humans have obviously gained a great deal from that revolution, but society can't stay on that path. Everyone now knows how human beings are contaminating the biosphere, but another troubling possibility has emerged: Humans will run short on easily accessible, clean biological and technical materials from which to build and create a beneficial civilization.

These ideas and concerns united us—Bill and Michael—as far back as 1991, when we met in New York.

We were united in a common value: "How can one design or manufacture in a way that loves all of the children, of all species, for all time?" We wanted our products to be a positive contribution not only to this generation of living creatures but to future generations, to the whole world.

We were weary of the finger-pointing we witnessed in environmental activism, which we felt drained human dignity and, frankly, slowed down what could be magnificent progress in designing a healthier world. So our coming together was the beginning of a profoundly inspiring friendship.

Not long after our first meeting, we were asked to create design principles for the 2000 World's Fair in Hannover, Germany, the theme of which was humanity, nature, and technology. So many people saw doom-and-gloom scenarios. We wanted the participants to focus not just on the aesthetics, efficiency, and utility of their designs but on the holistic quality and beauty of their design intentions, their relationship to the future and to far-flung places. The Hannover Principles were publicly presented in 1992, and they guide our work and thinking today:

1. Insist on the right of humanity and nature to coexist in a healthy, supportive, diverse, and sustainable condition.
2. Recognize interdependence.
3. Respect relationships between spirit and matter.
4. Accept responsibility for the consequences of design decisions upon human well-being, the viability of natural systems, and their right to coexist.
5. Create safe objects of long-term value.
6. Eliminate the concept of waste.
7. Rely on natural energy flows.
8. Understand the limitations of design.
9. Seek constant improvement by the sharing of knowledge.

Ten years after we articulated those principles, we wrote *Cradle to Cradle*, in which we further described how they could be put into practice. The book ranged across many design topics but focused primarily on products.

Which brings us to today, to this moment, with this book in your hands. *The Upcycle* asks you to reconceive broader aspects of the world, employing Cradle to Cradle design as a framework. It's time for us to begin restrategizing the design of our society as a whole. If our work over the past decade has taught us anything, it's that not just designers, not just architects or scientists have important ideas to contribute. To fashion a Cradle to Cradle world, we need everyone—the mothers, the fathers, the children, the teachers, the business executives, the politicians, the homemakers, the factory workers, the store owners, the customers, and so on—everyone.

In this book, we want you to consider design on all scales, from something as small as elemental carbon to something as big as the future; from something as basic as soil to something as extravagant as caviar; from not only how we design our world but how we power it. This is upcycling: taking Cradle to Cradle and applying it not just to how people design a carpet but how they design a home, a workplace, an industry, a city. Using the Cradle to Cradle framework, we can upcycle to talk about designing not just for health but for abundance, proliferation, delight. We can upcycle to talk about not how human industry can be just "less bad," but how it can be more good, an extraordinary positive in our world.

Start Where You Are

> Glance at the sun.
> See the moon and the stars.
> Gaze at the beauty of earth's greenings.
> Now,
> Think.
> —Hildegard of Bingen

When we came up with the goal statement for this book, *The Upcycle*, we did just as Hildegard, the 12th-century philosopher, theologian, and naturalist, advised. The two of us were tucked away with friends in a lodge in the northwest of Iceland, and as we worked on the wording of the statement, we looked around at the landscape and thought about what we saw: a wall of volcanic rock that the sun, burning low behind us, polished to burnt umber. A river, delightfully dense with Atlantic salmon, Arctic char, and trout, extending on either side. To our east, a green valley winding for miles toward the island's central glaciers. To our west, odd, imposing 50-foot-tall conical hummocks rising toward the sky with the silvery Arctic Ocean beyond.

This span of unique, diverse details awed us and reaffirmed the huge scale and intricate specificity of our creative challenge: to contribute our part to the bounty already present in the world and to humbly remember the mutability of even huge landscapes. Iceland had been forested in ancient times, and now the only trees in our view were those behind fences, protected by the farmers from the free-ranging sheep.

The sentence we came up with for our goal statement conveyed not how Cradle to Cradle worked, as we had described in the first book, but *why* it existed—which felt right to us, since the inspiration for Cradle to Cradle had always been the world we could help achieve, i.e., the upcycle of our daily work. It goes like this:

The goal of the upcycle is a delightfully diverse, safe, healthy, and just world with clean air, water, soil, and power—economically, equitably, ecologically, and elegantly enjoyed.

Why so many words? Perhaps only people so focused on abundance would describe the concept so verbosely. We wanted to honor and acknowledge everything the world can be and have within it if Cradle to Cradle design were truly to take hold. No limits. Instead, abundance: diverse, safe, healthy, clean, enjoyed. The upcycle is the opportunity to measure and develop the tools needed to meet this goal. It is important to realize this is always about constant movement and there is no finish line—no endgame.

Before we go on, though, we want to talk a bit about *how* we work. As we have said, we are fascinated by and excited about the atomic magic that can be engaged in the laboratory when inventing new ways to refabricate man-made products. But much of our work too involves recasting the very language society uses to define its challenges.

Let's take a simple example: Cradle to Cradle.

The first time you see that term, it might seem odd. Perhaps you might not even immediately grasp the meaning. But as it is repeated, that term can shift the very framework of how you think about objects. You might pause and begin to wonder, *Why is a manufacturer so proud to be responsible for a product from cradle to grave? What happens to the product after a consumer is done with it?*

Why do people even think of products as living things that go from cradle to grave? We know there is no grave; there are landfills or incinerators, where the product's components persist as debris, gases, and runoff and are lost for good. Oops.

Why have I accepted the idea of lifetime warranties for so long, when I should consider the need for a warranty after the "life"?

Why is the manufacturer not warranting that its product will be beneficial to the biosphere after the product's "death"?

By the time you have read the term "Cradle to Cradle" several times, you are in the headspace we would like you to occupy: questioning false beliefs. Free to imagine innovative solutions. Expecting more from industry, society, and yourself. That's where we want you and ourselves to be. And that's why you will find many instances in this book of us playing with language.

Now, when we say "play," we don't mean simply for fun. We want you to understand too that words can be "terms of art" with extremely specific meanings. We do not just talk, for example, about life-cycle assessments. We talk about defined biological and technical cycles (which we are about to describe for you). We want "Cradle to Cradle" to be an aspirational term, leading to constant improvement of a product or systems. So when you see a strange term we are coining, let it first open your thinking and then help you be more precise in what you expect from design strategies.

With that hope, we will now run you through a few key concepts from *Cradle to Cradle* so we are working with the same basic lexicon from the beginning. Let's start with those biological and technical food chains.

Why We See Abundance

One of the most important concepts in *Cradle to Cradle* is that materials can be designed to differentiate between the biosphere and technosphere and become nutrients forever. For example, the "waste" of an animal becomes nutrition for microbes, fungi, plants, trees, reptiles, mammals, and so on, perhaps even food for humans. This is a simple example of a *biological* nutrient cycle.

The term "technical nutrients," which we believe we coined, includes metals, plastics, and other materials not continuously created by the biosphere. Instead of these products becoming waste in a landfill, they could become "food" for another product, and that product would also become "food" again—endlessly.

To translate this to everyday use, think of juice boxes, the type of cartons used for many of the beverages people consume today. A typical juice box is an amalgam of aluminum, plastics, and raw paper that cannot easily be recycled (a very specialized and rare facility is required to separate them and re-form the material). Aluminum alone—a technical nutrient—can be recycled again and again at the same level of value, as long as it is pure. But if you add cardboard and plastics, you weaken the technical-nutrient quality of aluminum. The biological nutrient of the cardboard is tainted by the combination with aluminum. And what do you end up with? Mountains of suboptimal packaging in the dump or the incinerator. The precious aluminum is lost to its potential endless high-value cycle. The soil and air and water are contaminated.

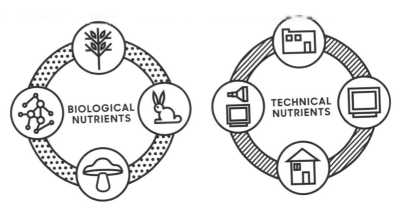

We call this conventional design "cradle to grave." It aims at only one use, period, after which the product and its materials are discarded, thrown away. But of course "away" can cause contamination of the biosphere. As Bill said many years ago, "Away has gone away." On top of that, society today wastes the *potential* benefit of these marvelous nutrients—all the useful products that they could become.

A "Regulation" Means "Here Is Something to Be Redesigned"

Where do we find the places to start rethinking our contributions, the opportunities to offer redesigns?

Silent Spring, Rachel Carson's landmark 1962 book about the effect of pesticides on the environment and particularly on songbirds, is credited with helping launch the Clean Water Act, the Clean Air Act, and the Environmental Protection Agency. Carson's lyrical, sorrowful homage to the diminishing diversity of the natural world— she imagined a world in which songbirds, frogs, and insects had disappeared—was a wake-up call to the seriousness of the design problem in pesticides. It spurred new thinking about how humans interact with nature.

Yet although regulations are obviously a valuable signal of concern by society—even vital at certain moments in human history—we can also consider them at some point to be alerts to design failures. Or, to put it more positively, signs of design opportunities.

PROPOSITION 65 WARNING: THE STATE OF CALIFORNIA REQUIRES THAT WE WARN YOU THAT THE PROPERTY CONTAINS CHEMICALS KNOWN TO THE STATE OF CALIFORNIA TO CAUSE CANCER, AND BIRTH DEFECTS, AND OTHER REPRODUCTIVE HARM. THESE CHEMICALS MAY BE CONTAINED IN EMISSIONS AND FUMES FROM BUILDING MATERIALS, PRODUCTS, AND MATERIALS USED TO MAINTAIN THE PROPERTY, AND EMISSIONS, FUMES, AND SMOKE FROM TENANT AND GUEST ACTIVITIES, INCLUDING BUT NOT LIMITED TO THE USE OF MOTOR VEHICLES AND TOBACCO PRODUCTS. THESE CHEMICALS MAY INCLUDE, BUT ARE NOT LIMITED TO, CARBON MONOXIDE, FORMALDEHYDE, TOBACCO SMOKE, UNLEADED GASOLINE, SOOTS, TARS, AND MINERAL OILS.

These words appear on an actual sign in a California public building. The government's need to regulate people's exposure to that building is a tacit acknowledgment that the building was not designed smartly. Had it been, then its interaction with people would be neutral, or even, in the upcycle, positive. The more time a person spent in it, the better.

The same goes for products with warning labels: A colleague of Michael's who recently moved to Germany told him that when she brought home her brand-new toaster, a warning label in the box stated that the toaster should be turned on a few times and allowed to heat up fully to burn off the coatings on the wires. The label asserted that the strange odors coming from the toaster at the beginning were normal and harmless—but that it should nevertheless be used in a well-ventilated area.

Now, is that good design? If a product were crafted to be healthful from the start, having more of it in the world would simply mean thriving businesses and satisfied customers. No one—not a businessperson, not an environmentalist—would want to stifle the product's creation, distribution, or consumption (or put it in a well-ventilated area). Regulations and warnings say, in other words: This thing exists, but it would be best for the health of humans and the planet if it did not. But, given that it does exist, here's how to minimize, though not eliminate, its awful effects.

When we realize the price we pay for careless design, it's clear that society might shift its thinking to consider good design not simply a luxury for the wealthy but a fundamental human right for everyone. The Declaration of Independence declares that human beings are born with the right to "life, liberty, and the pursuit of happiness." The United Nations has declared clean water and sanitation a human right. These things cannot exist without well-intentioned design. Designers do not have the right to inflict suboptimal design on all of us. Design is the first signal of human intention, and who would intend, who would purposefully set out, to design a system that pollutes our air, our water, our mother's milk with harmful chemicals?

Less Bad Is No Good

In *Cradle to Cradle*, we also wrote at length about people's problem of relying only on eco-efficiency to minimize their negative impact. Basically making a suboptimal system more efficient by curtailing how much "bad" it produces: not doing the right things, but doing the wrong things better. We might actually refer to this as "eco-insufficiency": It is insufficient as a strategy because it encourages us to stick with what is poorly designed—just to try to do less of it.

Recently, the scientific director at Michael's Environmental Protection and Encouragement Agency bought a name-brand juicer and was excited about having fresh homemade juice daily. To her dismay, when she made her first glass of juice, it tasted of chemicals. The Hamburg Umweltinstitut sent the juicer to a laboratory to test the level of off-gassing. The rates were shockingly high—more shocking, however, was that the juicer's off-gassing level was still within the legal range (up to 10 milligrams per square centimeter). That's an awfully good example of how eco-efficiency—setting a metric of tolerable levels of a gas—is simply not sufficient for positive design.

Eco-efficiency might also seek to curtail consumption: water use, for example. But that consumption limitation is premised on the notion that we live only in a world of scarcity and limit, that the ecological world is insufficient for the world of human activities and industries.

This is just not true.

Here's one simple example: Many people enjoy taking long, hot showers. In most households, when the water runs too long, one might assume that water and money and energy are being wasted. But if the water in one's house or hotel is filtered, recirculated, and solar-heated, people can luxuriously shower guilt-free for as long as they like. No one is worried that he or she needs to waste *less* water, energy, or money. The system is less—but it's more. The design is optimized around human nature and pleasure. It is purely beneficial and positive. That is eco-effective. That is our goal.

Since the publication of *Cradle to Cradle* 11 years ago, we have expanded our ideas, putting them into practice, understanding the obstacles to their adaptation, and trying to inch closer to a better-designed world. Back then, we only believed that such a world was possible. Now we have seen it becoming reality. We've seen a great deal in innovation. We understand whole new subtleties. And we want to tell you what has worked.

Cradle to Cradle, the book itself, was designed as a technical nutrient (and not, as some erroneously assumed, to be composted as a biological nutrient). It was made of waterproof synthetic fibers, inorganic fillers, and soy-based inks to be recycled as another synthetic paper product. It was not compostable by design. It signaled our intention to design for a human industry without waste, and it forwarded a strategy of hope.

The book did seem to inspire hope. But there has been one common reaction: Call it enchanted skepticism. Or engaged self-doubt. Many readers have felt that the articulation of a design framework for unique nutrient cycles—biological and technical—is a beautiful discovery. It's logical and even just great common sense. But how can an individual use this knowledge to make a better world?

How can I take action?

How can I, the small-business owner, embrace these concepts and still be competitive in the marketplace?

How can I, the customer, get companies to make such products?

How can I, the CEO or business executive, design and implement such a systemic overhaul and still meet customer demand and quarterly earning expectations—and how do I convince the board to go for it? This book attempts to answer those questions.

Since its publication, we have been delighted by the number of people who have worked to put *Cradle to Cradle* into practice in their own work. We have also come to realize just how ambitious our ideas were. At that time, few people had concocted the clean chemicals, devised the perfect joinery, or created the right dyes. Our job in many respects was and is to make that world exist, so other people will see that the larger vision is possible, plausible. When people buy a gorgeous fabric, for example, that during its manufacture left the water running out of the factory cleaner than the water flowing in, it gives them license to set similar goals.

The aftereffects of the projects we took on turned out to be larger than we dreamed. We have been astounded by the enthusiasm and the profitability that accompany taking up this thinking. Government regulations drop away when there are no ill effects to minimize. Cradle to Cradle designers and manufacturers know that they are engaged in what Buddhists call "right livelihood," a way of making a living within the framework of right behavior that allows them to happily present themselves to their children.

We are now convinced that people *can* do this. We have done this, and we work with major corporations and institutions to do this. Some of the world's largest companies became pioneering early integrators of Cradle to Cradle ideas, including Steelcase and Herman Miller, the international furniture manufacturers; Berkshire Hathaway's Shaw Industries, the world's largest carpet manufacturer; Ford, the iconic American car company; Cherokee, the environmental investment fund; and China's Goodbaby. In the past decade alone, we have helped hundreds of forward-thinking enterprises, from Aveda and Method to Construction Specialties, Delta Development Group, Desso, Gessner A6, Puma, and Royal Mosa, to name just a few. Many of our partners have made company-wide commitments to Cradle to Cradle principles. The ideas have proven vital not only to businesses but to governmental agencies, not-for-profit organizations, and communities in the United States, the EU, and Asia—we will look at some of their innovative initiatives throughout the book. Too many organizations and individuals have begun to upcycle to list here, but we feel it is worth underlining the widespread possibilities that come with Cradle to Cradle.

So our question is: Why don't you join us?

Our ideas may sound at times optimistic, even quixotic (we definitely want you to have fun expanding your thinking as you read), but we are working in the real world, with clients who are putting these ideas to tremendous effect every day. All these companies have visionaries on staff courageous enough to rethink the design of their products and systems so we can move closer to a Cradle to Cradle world. If they can do it, and enjoy the profitability, then you can too.

It's time to put away the scolding tone in urging industry toward more environmental thinking. We almost never find a CEO who doesn't want to make the company's product a known good in the world. The business community is interested in health and abundance. What most corporate leaders need is not chastisement but customers, encouragement, and support.

We tell them: We now *know* how to make your product healthy and safe for the environment. We now *know* you can power with solar, wind power, and other renewables. We *will* work together to get you the best technical nutrients coming out of other cycles into your factory. We understand that change might happen gradually, like the hatching of a butterfly. It might not happen in a flash. But it *is* happening. We can feel it all over the world. From the solar-powered campus of Google, to the roof farms of Brooklyn, to the phosphate reclamation in sewage treatment plants in the Netherlands, Canada, Japan, and elsewhere . . . we are watching the revolution unfold like a butterfly's wings.

From the beginning, Cradle to Cradle has honored commerce as the engine of growth and innovation, as *the* way to make the planet far more productive than it is right now. Suboptimal design cuts revenue and keeps businesses from being as profitable as they could be due to losses—in materials, in energy, and even in worker and customer health and enjoyment. Cradle to Cradle thinking can be business's innovation engine.

Consider this book an upcycle of the last book. Building on the ideas in *Cradle to Cradle*, we want to show you how people can move from being "less bad" to becoming part of the natural cycle of regeneration on the planet.

We can be overtly good.

We can finally enjoy our full human dignity. We can celebrate the unique and fruitful role we possess in perpetuating the biological system. We can proliferate. We can create more magical objects.

And we can, in fact, enjoy the satisfaction that a tree, a bee, the sun enjoys: While I exist, I make this world more fruitful.

Cradle to Cradle is a grounding and coherent foundation, the fulcrum against which we can lean our levers of desirable change. This book, *The Upcycle*, is an update and a collection of observations, evocations, and stories of continued improvement to be discovered the way the butterfly finds flowers in the garden. To us, upcycling is the most exciting project of all. It's going to take all of us. It's going to take forever. And that's the point.

Life
Upcycles

Chapter 1

What pictures come to mind when you think about the "environment"? A beautiful, pristine wilderness or an untouched, overgrown field? A rocky shore? Mountains rising in the distance?

Each of us has images that we associate with the word "environment": usually places that are "natural" and most likely unspoiled. Your thoughts probably flow to the atmosphere that surrounds and supports you: the air you breathe, the bodies of water you enjoy, the trees and grass growing around you, and the birds, flowers, and other living, nonhuman things that you sometimes observe and appreciate and don't want to see destroyed. This is one view of the environment: a sacred place we enter when we leave our towns, cities, societies, and machinery behind.

Ralph Waldo Emerson defined the scope of this ecstatic world in his 1836 essay "Nature": "Nature, in the common sense, refers to essences unchanged by man; space, the air, the river, the leaf."

For Emerson, nature was immutable, too big for humans to impact.

So much for the 19th century.

Clearly, humans have changed the air, the river, the leaf. Humans enter the picture and our arrival turns natural harmony into horror. We are the unruly children of a benign and generous Mother Nature. She gives us everything, and we, irresponsible and thoughtless offspring, make a mess wherever we go.

Our vandalism appears unlimited, in reach and in detail, on large and small scales. It is everywhere, all the time.

One piece of popular visual iconography at this point in history depicts polar bears wandering around looking for ice to walk on. As depressing as this sight is, under the surface, the molecular reality is perhaps even more disturbing. Polar bear blood now contains such chemicals as fire retardants used in babies' cribs 5,000 miles away.[1]

The view is that if we can't stop trammeling nature's riches, we'll bring on our own apocalypse. We wallow in the swarming, chaotic images of destruction and devastation that underscore our criminal behavior: global warming and melting ice caps; sea levels rising; ruthless floods, droughts, and storms; the pollution of waterways, oceans, and land from agriculture; toxic waste in landfills and on barges; piles and piles of unsorted offcast material; factory emissions; a strained "carrying capacity"; mining and mountaintop removal—not to mention social ills, such as child labor and unsafe workplaces, exploitation of resources by foreign industries, contamination of our indoor and outdoor air, starving people in the developing world, overpopulation, poisoned fish and dwindling stocks of deep-sea fish, a vanishing population of species around the world due to diminishing habitat (remember those polar bears) . . .

Stop! you might be thinking. *It's too much! We have to be less bad! We have to leave a smaller footprint! We have to make less of an impact, we have to get to zero—zero waste, zero emissions . . . Then things could be all right—if it's not too late!*

There are several flaws in this thinking about the environment. These flaws obstruct true progress.

We've just engaged in two of them.

1.　It is worth noting that in the last 10 years, PCB (polychlorinated biphenyl) levels in polar bear blood have dropped almost 60 percent. When society determined PCBs to be highly hazardous and requiring removal, the ban on their use had real effect. In other words, being less bad isn't bad. (It was good for the polar bears.) It is simply insufficient.

　　We want to move further to focus on what was substituted for the use of materials containing PCBs. That is the upcycle. As you will see on the chart on page 34, perhaps there is more to do than just reduce our negative content. Perhaps we can see opportunities to positively define future actions rather than defining them simply in the negative.

1. Romanticizing Mother Nature

We often romanticize Mother Nature, perhaps because so many of us have lost a deep connection to the larger world (what Richard Luov memorably called "nature-deficit disorder"), or because we feel sad or guilty about its depletion and destruction. We glorify the preindustrial past, the ways and pace of tribal cultures, and sometimes even wonder what it would be like to turn back the clock and live as those people once did.

But we forget: Mother Nature is no gentle lady. She can actually be quite nasty. In *The Wooing of Earth*, René Dubos points out that "the word 'wilderness' occurs approximately 300 times in the Bible, and all its meanings are derogatory."

Technology and experience have enabled us to mitigate some of nature's harshest treatment, but she can still be breathtakingly brutal. Naturally occurring botulinum toxin is 10 million times more potent than cyanide, gram for gram: What kind of mother would create such substances for her child?

The idea that nature is separate from humans, unspoiled and sacred, is well-meaning, because it attempts to make us think about our environment and its needs, but it often leads people to consider their own systems, their processes and products, as *removed* from nature. We pity this exotic, beautiful stranger, which leads to the next flaw.

2. Believing We Must Leave a Smaller Footprint
(Because We Are So Bad)

Once we have romanticized nature and begun thinking of ourselves as the unwanted child, the intruder, we do what any self-conscious, well-meaning outsider would do: We try to become less noticeable.

We cower, try to take up fewer square inches. We attempt to produce and consume less. We throw less "away," we turn food-grade water bottles into fleece jackets (more on that later), and we might even live more meagerly because we think this is the right thing to do. Numerous books and websites tell us how to minimize our carbon footprint, how to save the planet from disaster, how, essentially, to be less "bad."

Now, there is nothing evil in the desire to be less frivolous in using resources. Famously, Toyota, the icon of quality management, as well as thousands of companies reporting to the Dow Jones Index and the Global Reporting Initiative all over the world, have come out and said they have saved hundreds of thousands of dollars a year on energy and water, and that's their ecoprogram. These conservation measures are all valuable and well-intentioned exercises. But from a design perspective and an intentionality perspective, as well as from a simple business perspective, if a company touts its savings of hundreds of thousands, even millions, of dollars a year by using less water and energy, for example, for its production, any management consultant or expert would have to ask the fundamental question: If there was this much inefficiency in the system, money being spent on things that weren't producing value, was the management managing well?

Saving water and energy actually has less and less to do with being positively environmental. These measures have the benefit of being less damaging, but their actual implementation could have been done as a simple management process to reduce cost in the first place.[2]

Trouble often arises when a company is asked to go beyond eco-efficiency to paying for the license to pollute (emission permission) or being fined for the release of unsafe chemicals or materials. In the old model, companies that struggle just to meet their bottom line might reject what they consider the luxury of environmental

---

In chapter 4, we will tell you how environmental thinking can be oriented toward greater productivity and profitability, by using what we call the triple top line. Generative, not degenerative. Eco-effectiveness can actually create endless resourcefulness.

thinking. Outsiders respond by demanding more government oversight and control. Resentment builds on both sides. Blame and shame. It's an emotional quagmire, one that seems impossible to escape.

Ecologism, or How to Drain Pleasure from Life

These flaws in thinking and their attendant fear lead to something we call "ecologism." Ecologism refers to the strident metrics and mandates intended to "help" the environment that do not actually support ecologies or commerce. The reason for the word "ecologism" is that when any value takes precedence over all others, it gets an -*ism* added. Socialism. Capitalism. Ecologism.

Even though environmental efforts are often well-meaning, ecologism can be tyrannical: Its laws may only mandate that we save energy and water, minimizing the negative effects of poor design—in other words, "green-washing" the dirty laundry a bit.

Under this dictatorship of ecologism, we see more codes and standardization, more regulation that stunts economic growth and incentive, more limiting of consumer choices. Taken to extremes, in a world based on ecologism, only saving resources would matter and quality of life would be secondary. In a world of ecologism, taken to extremes:

- The cost of a plane ticket would vary according to your weight, since heavier passengers require more jet fuel to be burned than lighter ones do.
- A person driving a car could be fined for keeping the engine running at a red light.
- The preparation of spaghetti al dente would be mandated (even if you like your pasta soft and mushy) because that would save 20 percent on the power needed.

- Regular neckties might be outlawed in favor of bow ties, since bow ties require less fabric, which translates into a savings of dyes and water for their production.
- Everyone would have short hair to minimize shampoo consumption.
- Italian prosecco would be outlawed in favor of French champagne, because the smaller bubbles in champagne off-gas less carbon dioxide.

Think about attempting to fall in love less wastefully. Or what about an efficient child, or an efficient childhood? Terrible, right? Children, and childhood, can be—and we prefer them to be—full of richness, diverse enjoyments, fruitfulness, digressions, wanderings, imagination, and creativity. Who would want simply a "sustainable" marriage? Humans can certainly aspire to more than that. In all of life, people can think big.

No More Zero

Do not waste yourself in rejection, nor bark against the bad, but chant the beauty of the good. —Ralph Waldo Emerson

Imagine getting into a taxi and telling the driver: "I am not going to the airport." Is our modern environmental strategy simply to tell people what not to do? What if we are even being obviously unscientific?

A well-known car manufacturer recently ran an ad in a magazine that featured an image of a tree whose trunk was composed of people.

The caption stated that their goal was one commonly advertised by many companies in many industries: "Aim: Zero Emissions."

It's ironic that the car company chose to use a tree to illustrate its point, as a tree does not produce "zero emissions."

On the contrary, it produces emissions constantly. Positive ones. A tree not only emits oxygen (and some carbon dioxide) as it attempts to grow . . . it wants to get bigger. It wants to use more carbon dioxide, to emit more oxygen. Higher and higher emissions! Emissions that human beings like!

Yet many environmentalists and businesspeople want to eradicate emissions, because emissions have been labeled as bad. The International Air Transport Association, a trade group representing the world's airlines, has stated a goal of zero emissions by 2050. Other companies focus on micro-efforts to become carbon neutral, such as printing their annual or sustainability reports on recycled paper. Meanwhile, opportunities for more productive forms of optimization in their industries may go unnoticed.

When aiming for the negating, self-abasing goal of zero emissions, people rally to reduce population and curtail consumerism and development, wag fingers at industries, and keep us mired in the world of limits and poor self-image for our species as a whole. All this pits those concerned about the environment against those concerned about industry and rigidifies the division between "natural" and "human."

But there are 600 billion trees in the Amazon, all emitting oxygen. Imagine "minimizing" that or making their emissions "zero"! People would want no such thing. Humans breathe in oxygen, breathe out carbon dioxide. Trees breathe in carbon dioxide, breathe out oxygen. Natural. Symbiosis . . . not nothingness. In nature, emissions are as natural as breathing. Emissions *are* breathing.

Why should humans, with all of our brainpower and technology, be limited? Why should humans see ourselves as creating the only systems on the planet that can't participate delightfully and safely in the natural world, the way a tree or a forest or an earthworm or a colony of ants does?

The entomologist E. O. Wilson has estimated that the earth is home to between one quadrillion and ten quadrillion ants, across more than 12,000 different species—4,500 alone in the rain forest. Consider this quadrillion-to-billions comparison: the biomass of six-billion-plus humans compared to the biomass of at least a quadrillion ants (termites have the same biomass, but no one likes them, even though they are vegetarians). The ants' biomass is greater than ours. When humans are compared to ants, they don't even come close. At any given moment, the biomass, or stored energy, of the world's ant population is five times the biomass of our current human population, or roughly the equivalent of 34 billion people.[3]

The important fact is not just that ants have five times the biomass of humans. It's that, at that phenomenally large level of biomass, they have thrived on this planet and continue to thrive, redistributing their resources rather than (as people do) depleting them. Ants safely handle their own materials (what we might have called "wastes" in Cradle to Cradle) as nutrients in systems for their own and other species. They build homes out of recycled material. They grow and harvest their own food. They create their own medicines and disinfectants that are healthy, safe, and biodegradable.

We are not suggesting that humans live exactly like ants. That doesn't sound fun or inspiring. It sounds parsimonious (and, considering their practices of slave capture and child labor, unethical and regressive). But just because we don't want to live like ants doesn't mean we can't learn a great deal from their industrious

By "biomass," we mean the total amount of living matter in a habitat and the amount of energy, usually solar, stored in the individual organism. A creature has biomass whether alive or dead; the biomass of every organism differs one from the next, as does the total biomass of that organism, which is determined by multiplying its individual biomass by its population.

example. They do, after all, have full employment—all have jobs. If people modeled systems and communities on those of ants, they would not deplete resources, and they could flourish at a population figure much larger than they do now, making and consuming the things they want.

This is why we believe that a population of 10 billion living comfortably and fruitfully in a Cradle to Cradle world is not a pipe dream.

As Emerson wrote, "The vegetable life does not content itself with casting from the flower or the tree a single seed, but it fills the air and earth with a prodigality of seeds . . . All things betray the same calculated profusion."

In the 1970s, the economist E. F. Schumacher considered the possibility of optimized agricultural production and remarked on the complex and fruitful connectedness of grazing animals and honey locust trees, an interdependence that he had observed and translated to his own work. Honey locust leaf systems have a special lightness and transparency. Each leaf consists of a long stem in the middle, with tiny leaflets on either side. The leaves themselves are not opaque but filter sunlight so the rays can reach through the leaves to the ground, thus allowing grass and other ground covers to grow. Schumacher noticed that as sheep grazed on the grass and fallen seedpods under the trees, they left nitrogen-rich droppings that further fertilized the trees. The trees also captured nutrients from the atmosphere and sequestered helpful carbon into their root systems—the opposite of an automobile, for example, which releases carbon into the atmosphere to deleterious effect.

The grass was nourished for its growth by both sheep and trees. The honey locusts optimized the contributions of the sheep; the sheep optimized the contributions of the honey locusts; the leaves of the locusts optimized the recurrent solar income; the grass beneath the trees optimized all of the above; and so on.

This is a small but beautiful example of how humans might begin to transform their relationship to the planet: How can we make things with these sorts of rich, radiant, beneficial effects in mind? It takes careful looking, careful thinking, and careful doing.

Humans too can participate in a "calculated profusion"—a delightful, thoughtful, well-designed abundance.

We can look *up*.

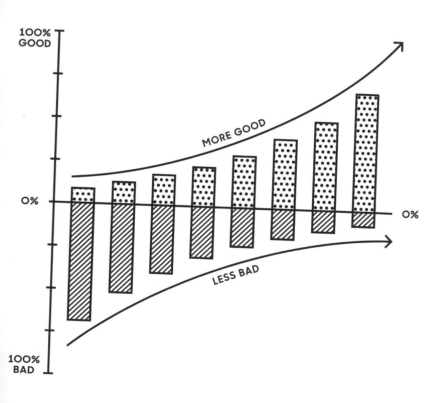

Trending Up: Getting Down to Business

We like the way businesspeople look at charts: A CEO prefers seeing an upward-climbing line. She wants growth. She wants increase. She wants top-line revenue for her managers to manage so she can have an ascending line of bottom-line profit.

Much of the well-intentioned environmental work thus far has preferred a different chart, one inclining down toward lower CO_2 emissions, decreasing population growth, fewer cubic tons of pollutants . . . What if those two charts, of the businessperson and the environmentalist, could be combined? Take a look.

The top side of the zero axis is the upcycle—striving to be "more good," not "less bad." Not only is the intention positive, but so are the numbers.

We know that the corporate world can understand why environmentalists have for so long focused on declining toward zero. How do we know? Workplace safety and health practices are often translated to charts that descend toward zero accidents—whether involving employees or the environment. The push to trend toward zero grew out of real concerns on the part of business leaders for health and safety, a desire to create an environment that encouraged employees to always be vigilant. In many companies with advanced safety protocols, workers are urged to act if they see unsafe conditions, with no fear of retribution for stepping outside of their described job. We honor that.

But what we now propose is that the desire to head to zero—no accidents, no spills—become not a culminating point on the graph but a midpoint or a crossing point. We prefer the ascending diagonal line to start below the zero and happily climb up to settle on zero accidents.

And then what?

What if that line kept on climbing, not just soaring toward zero accidents but pushed up and forward, toward better health for the employees, for the workplace, for the community in which the company does business? Take that ascending line even farther: What if it reached all the customers touched by the business? You could go from a workplace that does no harm to a health-giving workplace, to a health-*producing* workplace.

People need to think of where they might devise moments of optimization, of taking actions that improve employee life and the health of the community. What might those positive metrics look like? How about *more* solar energy produced per hour? *More* water purified during a manufacturing process? *More* tons of tomatoes and greens being grown on urban rooftops?

If human beings upcycle, they can all live productively on one planet. In other words, at this point in history, after so much damage has been done, people don't need to have less of a negative environmental footprint: They can have *a positive footprint*.

This allows us to come to terms with the human population "explosion" as a success story (how about a "population bloom" instead?). We could put forth a design model based on thriving people sharing the present with the future instead of running out of everything while toxifying the rest. What an opportunity, if humans live *up* to their potential, rather than living *down* in our self-imposed limitations!

Safe and generous abundance can become intentional and replace scarcity as the defining situation of our time.

Optimizing Life: Molecular Matters

Let's start at the very beginning, then, to see where this abundance might come from. Let's start with carbon.

This element is a gift to our species, certainly as important as oxygen and hydrogen when it comes to life on earth. For practical purposes, we can look at all living things, including ourselves, as carbon based. It is the element with the highest potential for making substances on the planet because it has four electrons on its outer sphere, allowing for four elegant connections to other molecules. It's the only element that can make these connections with such equality of strength, because the other elements with multiple electrons are too big or too small to do so.

All the amino acids require carbon. More than 95 percent of all the substances that humans actually need around us are built of carbon—water and metals, of course, excluded. Perhaps that is why so many people are anxious and even unhappy when they see things burning and lost to the atmosphere: We need carbon so desperately here, earthbound like us.

Carbon not only forms biological life, it is also incredibly valuable for plastics, pharmaceuticals, textiles, metal alloys, and much more. It is the key ingredient in fossil fuels such as coal, petroleum, and methane.

Despite being one of life's key building blocks, carbon makes up less than 0.02 percent of earth's crust. While humankind has been good at figuring out how to extract carbon, we have not been very good at devising a way to capture carbon once it's out, which means that the biosphere ends up losing it on a regular basis: car emissions going into the air, for example.

Of course, carbon in the air is natural at certain levels. Carbon is always being exchanged among earth's crust, oceans, soil, atmosphere, and biosphere. All these spheres essentially "breathe" carbon. They absorb and give back carbon molecules: Carbon dioxide, through the solar synthetic process, becomes organic carbon, for example.

Below earth's surface, a carbon bank exists that would not naturally work itself into the elegantly balanced carbon exchange. That bank is the fossil fuels. Fossil fuel is simply the decomposed plant and animal life from millions of years ago, pressurized and heated at tremendous temperatures. For now, consider it a bank of ancient biosphere.

All the talk about carbon emissions in the atmosphere and their effect on the climate has blinded us to an important reality. Commonly thought of as pollution, carbon emissions summon to mind images of filthy billows pouring out of factory smokestacks or the exhaust pipes of cars; but in fact what we are talking about is one of our most exquisite nutrients. As you will see in chapter 3, humans don't have an energy problem. Energy is abundant. What humans have is a materials-in-the-wrong-place problem.

People have taken crucial, valuable terrestrial carbon and put it where they can't reach it. Too much of it lingers in the air and water for it to cycle back at its natural rate. It would be like sending gold dust into the sky, or diamonds. Why did they do it? Don't they want it? Now what the heck do they do to get it back?

Of all the things humans might do with carbon, just about the worst is burning it—exactly what people have been doing for the past century. Fifty-two percent of the anthropogenic carbon dioxide generated since 1850 resides in our air (while roughly the other half is absorbed in our oceans, causing acidification). The carbon in the air has caused ambient temperatures to increase, the effects of which have been extrapolated by the scientific community to be leading toward future problems of flooding, severe and fatal weather events, and diminishment of species.

Environmentalists, government leaders, and even industrialists have talked of the benefits of "trading" carbon—giving companies a cap on carbon emissions that they may choose to buy from or sell to each other.

Now, turning carbon into a commodity can give well-meaning people who appreciate working with simple metrics certain checklists against which they can perform. That's all fine as long as there is a level playing field.

But carbon's real value is not in trade but in capture—as it has been for millions of years. It's a nutrient, not merely a numerical asset to be turned into a derivative and exchanged by traders in an abstract numerical universe.

The optimization of carbon is in soil, or in your children, or in your tomato. It's not "waste" to be minimized, and it certainly is too valuable to be dispersed the way it is now. In other words, human beings are burning up some of the most valuable capital, sequestering it out of normal ratios in the atmosphere and in oceans. What if instead we developed a high-carb diet for the planet?

Burning Man: The Uselessness of One Use

The other day we were considering the new rage for wood incinerators in Europe. Because the European Union set the goal of 20 percent power through renewables by 2020, burning wood pellets—small tablets the size of large vitamins—has become a government-rewarded form of generating so-called "renewable" energy substituting for fossil fuels. The idea is that instead of using ancient stores of carbon—which require drilling, cause dependence on foreign oil, create pollutants—people are regenerating biosphere carbon to create energy. They are not robbing our past (and future); they are exploiting our present. New trees planted to be harvested in a short growing period are meant to reabsorb the carbon released into the atmosphere. It's allegedly a closed cycle. Burning wood is even considered an "advanced

sidered an "advanced conversion technology" because incinerators take "waste"—in this case sometimes sawdust or reclaimed lumber or paper—and convert it into energy, into electricity.

It gets worse. Due to government subsidies, companies have realized that it's profitable to burn wood.

That might sound promising. However, Europe has a shortage of biomass—wood—so it imports most of its pellets from the United States and China. These pellet-burning plants release particulate matter into the air, not to mention higher levels of CO_2 than were originally projected. The air in these countries is likely to be getting dirtier, while more CO_2 is being released, even as the southern United States absorbs a bit more CO_2 because of all the rapid-growth trees that are being planted for harvest.

We found a label on a bag of wood pellets. Remember what we said about regulation being a marker of "something that needs to be redesigned"?

WARNING: COMBUSTION OF THIS PRODUCT RESULTS IN THE EMISSIONS OF CARBON MONOXIDE, SOOTS, AND OTHER COMBUSTIBLE BY-PRODUCTS, WHICH ARE KNOWN TO THE STATE OF CALIFORNIA TO CAUSE CANCER AND REPRODUCTIVE TOXICITY. THIS PRODUCT IS NOT FOR HUMAN OR ANIMAL CONSUMPTION.

Something needs to be redesigned.

In addition to the toxicity issue they introduce, the pellets have to be compressed into that handy form—refabrication that takes energy. It might be only 10 percent of the energy from the biomass harvested, but it is still a "line loss," energy whispering away.

Add in the carbon expended for shipping the pellets from the United States to Europe, or China to Europe, and we wonder if this is the best solution. Sure, you aren't robbing earth of its finite ancient carbon stock, but you are dealing with a whole host of problems that seem old to us, including pollutants. It sounds good only because it helps countries hit the metric of 20 percent "sustainable" energy by 2020.

Making the Most . . . of Everything

In this new model of energy production, the tree is chopped down, converted into pellets, and incinerated. Perhaps the tree detours briefly into an incarnation as furniture or paper before it's burned. You can definitely say that pellet incinerators have created energy from wood, but we contend that the incinerators have also wasted ample opportunities for abundance, and for jobs—the endless resourcefulness of materials, energy, and people's creativity—or, literally, "woodworking."

A pellet incinerator advocate might argue, "Well, some of the wood did become paper or lumber. It did have one more round of use." Too often, such wood gets treated with chemicals—glues, or copper in the case of pressure-treated lumber—and the residues result in ash full of toxins.

The even deeper problem, though, is that the process has robbed the wood of its potential for multiple uses. We probably don't need to argue too much over whether multiple uses are better than single uses. Human beings know instinctively that loss of opportunity is acutely painful.

Why, across cultures, is the death of a child so tragic? Because that child never had the opportunity to flourish in all the ways a human flourishes—to marvel at the world's natural wonders; to experience the joy of perfecting a craft, a sport, an education; to fall in love and dance the night away; to read a novel that shakes one's soul; to be a father or mother, a friend, a source of wisdom for the next generation. Sure, the result may be the same: We all go back to the soil. But what is lost when that life is cut short?

Opportunity. Abundance.

Although obviously not on the same scale of tragedy, the loss of a tree's permutations is regrettable too. If one considers every element in this world for its potential yield, think how much more fruitful the world would seem—and how many more people would benefit—if humans allowed the yield to unfold.

Let's start from the top. The tree stands before us. That tree— that particular use of carbon—is a great little machine. Imagine that you needed to design a machine to clean the air. How much time, material, and false starts would it take for you to fabricate such a machine?

Or if a company had crafted an amazing machine to clean the air and circulate oxygen, how much money would you pay for it? How much would an air-cleaning machine cost that lasts a few decades or maybe even a few thousand years?

And that air-cleaning machine also provides shade, animal habitat, food . . .

Now do you value the tree more?

Right now, the tree is free. It doesn't ask anything of you for it to do its job, nothing except not to be cut down.

We're not saying we should never cut down trees. Humans need trees for building key objects in our lives. Wood is an amazingly durable, tough, and beautiful material that, if nothing toxic is added, will biodegrade right back into the soil, returning all its carbon and nutrients. People will always need trees for their uses. But we think we need to be more intelligent in how we design these uses.

Here's a better life cycle for the carbon in that tree. What if, when you cut down the tree, you designed a table, not just for its first use but also for its endless reuse, its endless resourcefulness? What

Closing Some Loops

We've learned a great deal over these years about where we need to adapt and clarify our language and even to evolve our thinking, such as when we realized the fundamental notion of "Respect diversity" would be better as "Celebrate diversity." In other cases, we have learned how people relating to Cradle to Cradle can misinterpret our terms. Our statement "Waste equals food," for example, has been misinterpreted as simply meaning closed loops. That is why we now prefer to use the term "nutrient management," as described earlier.

Cradle to Cradle is a richly informed engagement with the biosphere and technosphere within continuous use periods over time. It can, on occasion, be as simple as putting a carpet through basic continuous use periods; or oftentimes it is complex, such as in our concepts for leasing products of service, the Materials Bank, or the Intelligent Materials Pool.

When we first proposed biological and technical cycles, based on Michael's intelligent product system as he conceived of it in the early 1980s, we called for the positive definition of product chemicals and components down to the smallest parts. We pointed out that most materials are not recycled but downcycled—degraded in quality through the "recycling" process. In terms of manufacturers, their misinterpretation might be to create products that dispense with the "grave" part of the cradle-to-grave linear conception of a product. This leads to a two-dimensional version of a closed loop—and potentially a harmful one.

For example, a manufacturer will work to make sure that a product's materials can be melted down, mashed, or otherwise processed into materials for reuse—for the same product and for other products. This can allow a company to be highly efficient—to seemingly cut its need for new sources for its manufacturing process and also, possibly, decrease its dispersal of product "waste."

But closing the loops in this way isn't Cradle to Cradle. A closed loop is not a positive event if the material being reprocessed is toxic. As just one example, if a carpet is originally manufactured with soft PVC, then, at the end of its first use period, melting down the carpet and recirculating it through the manufacturing process to make new carpets is far from ideal or optimized. Soft PVC is not a positively defined material (and certainly is a poor choice for carpets, which degrade over time and release particles into the air).

It's a "wrong" material. It's not appropriate.

In fact, ironically, PVC was initially put into mass production as a way to close loops in the manufacture of paper and hydrocarbons. Those processes created huge amounts of chlorine residue of sodium hydroxide and acetylene. Industry chose to put those by-products into PVC, a choice that demonstrates how modeling manufacturing on a simplified version of a closed loop is often not sufficient. (Because hard PVC is manufactured without the problematic plasticizers or UV stabilizers found in soft PVC, we may have to consider sequestering these products in hard PVC piping deployed underground for the next thousand years or so.)

In other words, if you close the loops on an existing suboptimal design, then you're not truly Cradle to Cradle. You have not optimized in three dimensions.

An outdoor-shoe manufacturer has been doing what it calls closing the loops on its shoes without using positively defined materials. It's admirable that companies are inspired to try to be Cradle to Cradle in their manufacturing and production processes—in the upcycling spirit, a first step in the right direction always deserves celebration—but closed loops without positively defined materials don't cut it in the end. The solution is not sufficient.

When making a plastic bottle into a fleece garment, you haven't upcycled. From the materials perspective, a plastic that was food grade is no longer food grade. A plastic that may contain antimony has now been attached to people's skin, and dyes and fixatives and rinses have been added to it. Attach a nylon zipper and metal snaps to the fleece and, in effect, the material itself has lost the capacity for recycling because it is mixed with these other things.

How could this bottle-to-fleece process be upcycled instead? If a company wants to upcycle polyester from a bottle to a fleece and it doesn't yet have polyester produced without antimony, it can rinse the polyester in an acid wash to remove the antimony, shred it, and put it through a bath. This traditional polyester, which includes heavy metals, can in essence undergo a purification ritual that removes the contaminants, leaving the resulting polyester toxin-free. At that point, instead of adding undefined colorants and rinses and fixatives in factories far away that do not have coherent protocols in place, this fleece, with clean chemistry, might have polyester buttons and threads added, which would make it very simple to recycle.

if you made a fine, sturdy table, augmenting the joints with glues that were formulated to seep safely right back into soil? Or you used metal joints, simple clamps that could be easily removed and sent through a technological cycle, while the wood went on its biological path? Or you used no metals at all; the table is designed to be fastened with wood itself? The table functioned for many years, maybe for many generations. One day it broke or fell out of fashion. The table was ready for its next cycle.

This time, because the wood has not been tainted with toxins, it could be made into particleboard, again using only glues that could go directly back into the soil. From there, the particleboard could be made into paper or insulation pulp, and then, when those materials were fully used, it could be burned, making the same amount of energy as it would have if burned as a tree. Because no materials from the technical cycle had been introduced along its path, the result would be a healthy ash that could feed back into the soil's nutrient load.

By letting the tree work its way through the system—in a clean loop—not only do you get all of those new uses for the wood, you also have 40 times more jobs: the carpenters, pulp makers, paper makers, and so on.

This is a cascading effect in terms of the value of the product.

At one point when we were considering this multiple life span of the tree, we wondered: *Might this not be called downcycling?* The tree is going from a table to maybe toilet paper. Isn't that a downcycle—even if you look at it just in crude terms, something for which you pay hundreds of dollars versus something worth pennies?

But if the carbon molecule is passed on each time in a manner that makes it easily reclaimed by the breathing of carbon in the atmosphere, biosphere, and so on, then it is not a downcycle. Each item is just passing the carbon molecule on. If it gets back to soil or atmosphere, it gets reorganized by the photosynthetic act. Isn't that beautiful?

Every life creates more opportunities—is beneficial—for the next lives. As long as the carbon becomes part of the soil in the end and is not contaminated by materials that toxify living systems, effectively it can't be considered downcycled. When we thought about it, we thought of an even more positive spin (why the negatives?). Left to its own devices, life always upcycles.

As Francis Crick pointed out in *Of Molecules and Men*, life likes to grow, take advantage of free energy from the sun, and operate in an open metabolism of chemicals for the benefit of organisms and their reproduction. More, better, endless resourcefulness. The upcycle.

This is so simple a notion—we realize it is essentially the building block of our work to date—but it has such elegance and resonance. A pessimist might argue that human life *is* a downcycle. You reach your peak of strength or fecundity and then you begin the downcycle to death. But we don't buy that. That is looking at only two metrics for life.

What about the information you accrue over life that can be converted into solutions? What about the contacts you make that can be converted into action? What about the wisdom you gain that can be converted into pleasure? Or friends into a web of community? Humans can begin celebrating our own growth, our own "emissions" over our lifetimes, to engage in a fruitful way with the rest of the natural world.

So we thought this was a revelation: You cannot downcycle life. Life upcycles.

It is valuable at every step. The insect is not lowly compared to the human. It's just all part of the web of life. So there is no such thing as a waste. No human who's a waste. No tree that's a waste. No piece of toilet paper. Upcycling eliminates the concept of waste.

Tending Our Terrain: Not Us vs. Them

What is, as far as we can tell, a key difference between us and all other species: our intentionality.

An ant can intelligently pursue strategies for optimizing its survival and that of its species. It can even inadvertently help other species thrive. But we do not expect that it can conceive more broadly of systems beneficial for species beyond its own. We humans have the ability to see beyond our species, and that ability confers a responsibility.

A tool's actual value is the purpose to which it is put. A hammer is terrifically helpful if driving a nail but diabolical if used as a weapon. The tool's intended use ultimately defines its value in every context.

We humans can define and describe ourselves by our intentionality, by the way we use intention to grow biological systems and maintain technical systems. We like what Claude Lévi-Strauss wrote about communities:

Cities have often been likened to symphonies and poems, and the comparison seems to me a perfectly natural one. They are, in fact, objects of the same kind. The city may even be rated higher since it stands at the point where nature and artifice meet. A city is a congestion of animals whose biological history is enclosed within boundaries, and yet every conscious and rational act on the part of these creatures helps to shape the city's eventual character. By its form as by the manner of its birth, the city has elements at once of biological procreation, organic evolution, and aesthetic creation. It is both a natural object and a thing to be cultivated; individual and group; something lived and something dreamed. It is the human invention, par excellence.

If the city does indeed occupy that space between natural object and artifice, then humans need to think about what their contributions—their artifice—say about their values as we interact with the natural world.

Let's look at one example: The rain forest often represents the epitome of untouched nature, a fragile prey of human avarice. The lamentations abound about how people are destroying this home to great biodiversity, trees that clean our air, a last refuge of unsullied paradise. Humans need to get out, the sentries of Mother Nature say, and leave the rain forest alone.

And even though many poorly designed ideas have wreaked havoc in the rain forest, and continue to wreak havoc in the rain forest and similarly critical ecosystems, people need to realize they can have—and historically have had—a more symbiotic relationship.

The science journalist Charles Mann, in his book *1491: New Revelations of the Americas Before Columbus*, points out that the Amazon rain forest can be seen as a garden, tended by humans and encouraged to grow in certain ways: Amazon tribes would often select little saplings in their journeys and hunting forays that they wished to promote and nip surrounding seedlings in the bud. If they preferred a Brazil nut tree to another kind of tree, they would snap the leaves off the other tree, or snap the head off the other saplings, allowing more space for the Brazil nut tree, which nourished the tribes, to flourish. These people engaged in tiny acts with consequences after millennia. They recognized that the forest was capital, not just currency, for their tribe. They anticipated the needs of the people who would come after them—and laid the groundwork for an enriched place. The rain forest is a garden. Humans used intention to tend it.

Humans can devise enriching systems by preference. They can take opportunities to interact with the rest of the biosphere in a mode of growth and abundance. We all need to get creative.

In *Cradle to Cradle*, we discussed the idea of designing a building like a tree. But what if that building offered not only shade, oxygen, and such, but also an abundance of food for the people in it? Factories, office buildings, schools, railway stations that also have giant greenhouses on the roofs. What if our railway rights-of-way were lined with solar panels as service products leased to generate energy? What if our doorknobs leached vitamins and minerals we need instead of elements disruptive to human biological systems? We will discuss all of these ideas in this book.

With the upcycle, it's clear that we intend that things get better for generations in the future, not worse. Why not, like the tribes in the Amazon, perpetuate those things we like, that are useful, pleasurable, and healthy, for millennia?

Let's take this even further. People can think beyond materials and energy to communities, ecosystems, cities, the planet. What if, as in some Hindu conceptions, every time a human being imagined the next state of existence, life got better? How can humans—the people of this generation—upcycle for future generations? How can people pass the carbon and water molecules along, benefiting all the generations that will follow us? How can people love all of the children, of all species, for all time?

It is *the* question, we believe. No matter the intricacies of the professional challenges we are facing, or the particulars of the project we're in the midst of, this question focuses us. It's the great one we aspire to respond to with our own work.

Houston, We Have a Solution

Chapter 2

Recently, Bill attended an innovations summit for one of the world's biggest apparel makers. One after the other, brilliant people took the stage to talk about the fabulous fabrics they had engineered to ignite consumer imagination. Who could fail to be impressed? Wizards from MIT and London Fashion This and Paris Textile That described shirts that could light up. Clothing manufacturers demonstrated materials, originally designed for military application, that could recharge batteries. New Age designers had made dresses that glow with your emotion. Gee-whiz inventors grew bacteria into membranes that became jackets.

It was all interesting, of course. But when Bill closed the conference for the senior management, summarizing the day's events, he posed the question: What's next?

He wasn't just asking what was going to be next year's coolest, most technically amazing, or most fashionable idea. The question, fundamental to the Cradle to Cradle nutrient reuse concept, is: Can this garment return safely to the technosphere or biosphere without contaminating either, while being rejuvenated as materials for new cycles? What happens to that glowing dress *next year* after being *this year's* coolest fashion?

What's next?

How will that luminous material being designed become not only a fashion magazine cover story but also a rag and eventually soil? Or, if not soil, a technical object designed for and maintained in the technosphere, made back into useful objects of the same high-quality material, used again and again without entering or contaminating the wearer or the biosphere: a polyester T-shirt (no antimony!), for example, which becomes a fleece, which becomes a safe polyester toy, which becomes clothing, which becomes containers, which becomes clothing . . . in the continuous sequence of its next becoming.

If technical nutrients aren't being maintained for their reuse cycle, they are commonly smashed together into monstrous hybrids. They get dumped into the biosphere. As soon as you let the next big thing into the biosphere, it's a neurotoxin. Whoops.

Use What's New: The Ingredients for Possible Solutions Are Always Being Invented

We don't mean to imply that thinking about design in this way is always easy, whether it is building design or chemical design or any other intentional act. Often people become so preoccupied with making an object "work" in its first cycle that they don't look at the next picture.

But let's take a simple example of how people's minds encompass this way of thinking every day: cooking food.

People don't think only about how the meal "works" as food— that it tastes like food and looks like food. They think of its future function too, if it will provide the needed nutrition and calories for future activity. When they choose the ingredients at the supermarket, they are already conceiving of the materials that will make them feel better, run better, work better.

People can train their creative thinking in their work for this kind of practice too.

But first you need to open up your imagination.

Let's look at an example of an innovator at work. Paul MacCready, an aeronautical engineer who passed away in 2007, cocreated the first totally human-powered airplane. When Bill asked how MacCready and his design partner, Peter Lissaman, were able to come up with the perfect solution, MacCready said he used an approach that involved stepping outside *what had been done* to more freely imagine *what could be.*

The flying machine they invented, the Gossamer Condor, won the prestigious British Kremer Prize in 1977, awarded because it was the first device designed to harness the energy potential of a human being to loft a plane a mile in a figure-eight pattern. When MacCready was asked how he was able to achieve this feat, he said he had three frame conditions for design.

1. Need

MacCready didn't personally need to fly a figure eight, a mile in total distance, but he desperately needed money. MacCready had cosigned a friend's loan on a venture that went bankrupt and owed exactly $100,000. The Kremer Prize, which had been offered by the industrialist Henry Kremer to the first person to solve the problem of human-powered flight, paid out £50,000. When MacCready noticed in the newspaper that the exchange rate had hit $2 to £1, he became more interested in human-powered flight.

In other words, incentive was present, and very often for inventors, it's financial. This isn't always the case: Many entrepreneurs are motivated by a passion without concern for risk. But frequently inventing is a highly incentivized exercise of some form of need. As is famously said, necessity is the mother of invention. But it's not the *only* mother. It might be better said that necessity is *a* mother of invention. If you are trying to survive in the woods, you might want to look under a rock for a spring. You might create a lever.

2. Imagination

To design the craft, MacCready and Lissaman—freely and unafraid—needed to imagine that they had never seen an airplane before. If they thought about conventional modern models of aircraft, they might get stuck with short wings or, looking back, find a biplane, which would provide inadequate lift and add excessive weight.

MacCready had to think about the human not simply as a pilot or a passenger but as an engine. Since the largest muscles in the body are in the legs, the pilot was a bicyclist.

The bicyclist would need wings that were so light they could lift him and themselves. The airplane would essentially become an organism composed of a bicyclist and materials. Central, compact musculature with long, light, feathery arms . . . like a bird.

3. Materials

MacCready wanted a material for the wings that was extremely strong and nearly weightless. We find it interesting that one of the things that inspired him was his memory of butterflies on the Connecticut shore of his childhood, when he would marvel at how they flew so well with delicate wings (we will talk more about the butterfly later).

For MacCready to solve his problem, he didn't rule out any material, no matter its cost or his previous experience with it. He had to be free to explore, to go wherever he needed to go to seek solutions. What was his solution? Mylar, a strong, sophisticated advanced polymer film, first used in the 1960s by NASA for a satellite balloon.

MacCready didn't reject Mylar because it had never been used for an airplane's wings, nor did he reject it because it's expensive. To solve his problem, he went wherever he needed to go, even to this relatively new material.

This, from our experience, is crucial. You need to solve your problem using whatever is available to you, no matter the cost or the rarity. You can always return to the reality of schedules and budgets when the time comes to build.

Thomas Edison's work exemplifies this notion. He wanted a natural substance for his lightbulb filament and even tried human hair but finally suspected that bamboo coated with carbon would work. To find the exact bamboo for the job, he sent explorers to Japan and China. He didn't think that paying for botanists to comb the forests on the other side of the world was too much to invest in solving the problem. He correctly surmised that if he were to discover a workable solution, investors and companies would be only too happy to take over the issue of reducing the cost of importing that bamboo.

We are not suggesting that you fritter away money on your answer. But you need to come up with a solution first. Once you have solved your problem, you can deal with the issue of cost. Either you can invent a cheaper way to make the same thing, or the market demand will begin to reduce cost, or you can look at the bigger picture to see what cost this slightly higher expenditure might be offsetting. A generous maker of a given material might even be interested in seeing your application in practice . . . because it might lead him or her to a larger market.

Don't let fears or restrictions get in the way of actually discovering a workable solution. You can work backward later.

Here was MacCready's beautiful solution:

The plane flew 1.3 miles at a leisurely 11 miles per hour, but it required only 0.35 horsepower.

If we take you to the big picture now, MacCready's conditions are immensely useful to keep in mind. For example, people *know* they have a defining need to replace fossil fuels and nonrenewable resources with safe, healthy, clean power and uncontaminated technical and biological nutrients. They *know* they need to find new ways to feed hungry populations. These needs are tremendous and will only grow.

Second, as we have already said, people can step outside convention: Imagine systems without limitation and without being trapped in old design paradigms, as if the object or system had not been made before. In this way, human beings can conceive of something more beautiful, more fruitful.

Third, people can consider materials and resources that haven't been used before, or consider old materials in different ways.

In the context of the upcycle, these three strategic conditions are very interesting for inventiveness, but they don't describe underlying "values" relative to the quality of the result, especially in the context of what happens to these things next.

MacCready solved several what's-next problems for our future that can be healthful for our world, including how little energy might be required to fly a plane if the right design and materials are employed. That was the question he wanted to solve.

MacCready's one-of-a-kind plane would clearly become a relic—in fact, a reliquary in a museum. On the other hand, if you're inventing a product designed for production, which would be liberally distributed around the country or across the world, like a plastic bottle, then "What's next?" becomes a very important question. As one sees with advanced polymer bottles today, what's next has been questioned. What if the bottles are going to be in the ocean, or burned? How will they be reused or recycled? These questions become fundamental to the core design of a product meant to be manufactured in quantity for the marketplace.

Let's stop for a minute and consider the lightbulb, 100 years after its invention. Lightbulbs solve for the currency issue of supplying illumination.

The filament moved on from carbon-coated bamboo to tungsten in the early 1900s. Recently, tungsten bulbs were replaced by compact fluorescents, supposedly society's salvation because they burn less energy. The curly compact fluorescent lightbulb is much more efficient in terms of its use of power than the conventional lightbulb; a lot of us have these in our businesses and homes, and may think it is terrific that the burden on power plants and our out-of-pocket costs for our electricity have been reduced.

At the same time, these bulbs have mercury in them— highly toxic, and the cause of a scary situation if one breaks on the floor of your child's bedroom. One problem has been solved while another has been created. Obviously, that can't be an effective path to clean power. If you make lightbulbs that are more energy "efficient" but make them out of toxic substances, how much progress is that?

Part of what we are explaining with the upcycle is that we can add a whole other dimension to your consideration, to this inventiveness. We can ask the question: What's next?

Landing on Earth: Design for Five Planets

Some people have estimated that earth's carrying capacity is limited to about two billion—that is, if humans continue to exist and grow the population at present rates, they might need five earths to support everyone (assuming the world trends to United States patterns of use and consumption). Some scientists have even suggested that as a species we should be aspirational by literally looking up, to planets beyond ours, in order to colonize.

We like to say that if designers only designed better, humans could support the 10 billion souls we expect will inherit the earth by midcentury (whether we like it or not). Perhaps we ought to honor the life of every child born rather than bemoaning their existence.

Rather than go up and away from the planet in order to save ourselves, let's look again, freshly, at where we stand. Instead of a moon shot, we need an earth shot. Imagine that we are coming to earth from elsewhere. What would it mean to arrive, to gaze around, knowing that we are going to live here. Then to *think*: "This is where we are. This place. How can we optimize it?"

Bill recently proposed this idea and designed and built it in California, at NASA's Ames Research Center in Silicon Valley. The building opened on Earth Day 2012.

The project began when NASA approached Bill to see if he wanted to help with the concept for the future Mars space station. They needed a structure that would perfectly coincide with Mars's environment, sustain human life, and use the resources readily available on that planet without demanding large imports of energy or water or any other requisite material. They wanted a livable structure that would keep Mars intact as humans thrived. They asked Bill if he was interested.

Bill said, "That sounds very exciting, but why don't we try something else first?"

He explained: It's not fair to simply make your revelations far-flung when you could bring them down to earth, when you could share them with us right here in real time. Why think only about colonizing other planets when humans still don't seem to know how to construct fully sustaining high-tech buildings on earth?

NASA could see the reasoning. Of course, we could learn a lot about building on earth by pondering an international space station for Mars, a very extreme environment. We also could learn a lot about Mars if we reflected back on supporting human life here on earth. These two things could inform each other tremendously and, again, prove to society the enormous value of scientific exploration for the betterment of the human condition.

Bill and NASA convened an early meeting of key participants in the room at the Johnson Space Center where Mission Control heard *Apollo 13* pilot Jack Swigert's famous words, "Houston, we've had a problem here." Bill felt that it would be inspiring for everyone to start imagining being able to say out loud, "Houston, we have a solution."

At that meeting, Bill met Dr. Steve Zornetzer, the NASA scientist who shared the vision and would become the leader of the project.

The result is a beyond-state-of-the-art research and development facility called Sustainability Base—a name echoing and honoring Tranquility Base, site of the first moon landing. It has been characterized as the highest-performing building in the federal government and was built on schedule within a normal federal budget. Of course, because of its experimental nature, research into its operations is ongoing. Humbly structured for continuous quality improvement, the building can, intentionally, be perfect by being imperfect. The building is a compelling, inspired signal of human intention coming from our most advanced scientists and engineers, who fully enjoy the fact that the work of progress is, by definition, always a work in progress.

When Bill and the team started thinking about the project, they tried what Paul MacCready did: They began without preconceived notions. They imagined they had never seen earth before, never seen a house, or an office building. They imagined they were beings from another planet, very much like humans, landing in Moffett Field in 2011. These beings have approached our planet with good intentions, and they have some of the same needs that humans do for air, water, energy, warmth, and coolness.

Stepping out of their spaceship, they immediately delight in the hospitable climate that is the earth's (not to mention Silicon Valley's). The place in front of them looks ideal for a research facility. How to power such a structure?

Just look up: the sun. It's right there, a thermonuclear reactor, exuding powerful photons that arrive wirelessly in eight minutes (NASA invented the photovoltaic cell for the purpose of capturing this energy in space to power satellites). No reason to fuss with messy alternatives, such as building an expensive nuclear reactor; burning the trees; or even digging a hole looking for ancient sequestered solar income, and burning it. Who would consider these complicated alternatives while a furnace of such brilliance pours its rays down on us? Problem solved.

Our imaginary visitors now wonder: *How do we keep this building from getting too hot?* Again, no problem. Just look down: the ground. Just as the sky provides heat, the ground provides coolness—a perfect 55° Fahrenheit. The visitors can sense, and even feel, a depth of coolness in the ground, even as they enjoy the warmth of the sun's rays. *What a magnificent place to build*, they are thinking. *It provides us with everything we need!*

Now, where would the building's water come from? It doesn't take long before the visitors detect moisture in the air, feel a light rain on their skin, and realize that the sky provides moisture, cycling through bodies of water on the planet and through all of its systems. Clearly, clean water is of great value and is part of a beautiful and supremely elegant system. Toilets and urinals use rainwater and gray water from sinks and showers for flushing. The gray water can also be filtered for irrigation of the surrounding vegetation. Why use water once when the more the merrier?

The building, in the future, can accommodate thinking about solid human waste not as sewage but as nutrients for the plants and trees. Or the solid wastes might one day be treated as a larger-scale nutrient system. They would go to nutrient systems installed by cities (more on that in the next chapter). But that's for another day. The building's system can and will be upcycled.

The 50,000-square-foot building uses what Bill calls "artifactual lighting." It's still artifice because humans had to intervene to create the window or skylight, but it uses an advanced daylighting technique. The people working in the building will only need to turn on the artificial lighting 40 days of the year, most commonly in the winter, when the hours of daylight are reduced.

One might expect such a high-tech building to be extremely costly, but all the base building systems were produced within the normal budget set for a federal building. In keeping with NASA's guidelines, special systems that the design group wished to incorporate had to be offset by equal monetary savings (for example, in energy consumption) so the extra investment could be recouped within a seven-to-ten-year payback period. Those extras totaled only 6 percent of the budget, all with documentable payback periods.

One exciting addition is a highly sophisticated "self-monitoring" system, akin to the systems used in spaceflight for the craft—in this case, the building—to read and predict all sorts of energy needs. For example, the system might understand that a weather front about to come through will cool the ambient temperatures of the building, and so the system does not need to power up cooling for the large meeting that just convened with the heat generated by more bodies and laptops. Energy saved. Energy use optimized with local nature.

To think about what it takes to upcycle a building, imagine driving a car without a steering wheel, an accelerator, or a brake, and no other data. There are many buildings in the world that are like a car without a gas gauge or speedometer. Kilowatt-hours come in and are used and you never know where they came from or what they are being used for. No indication of direction. No ability to signal intention, as with a turn signal. No awareness of consumption of fuel in the context of what's available.

In the average American house, people set the thermostat and don't open windows. Whatever the cost to maintain that condition, it costs. There is no optimization occurring in terms of fuel use or anything else. To decide when to open the windows requires a person to take the time to notice conditions. With pollen, and traffic noise, and air-conditioner compressor noise, it's easier not to worry about it. But it means that people have become mindless.

A system of louvered windows allows wind to pass through the NASA building, to refresh the air and precool the building for the next day. (NASA scientists study climate, and they know the weather. They have astonishing data on the upcoming weather, so they can anticipate the building's thermal characteristics over the next few days with great precision.)

Design could optimize systems and then let humans have the control. At NASA, although the windows can open and close automatically, humans can simply open a window the old-fashioned way. Our thinking was, *What if someone wants to hear the song of the bird that just landed on the tree near the window?*

If it's cost-effective, why wouldn't people want a system that can accommodate what an Aristotelian engineer might consider a Platonic search for truth and beauty in the surprise and delight of a richly experienced natural world?

Don't Limit Yourself with Your Goals

So we have run you through human-powered flight, the lightbulb, and an earthbound space station. What do these examples have in common? People who, while searching for a solution, refused the concept of failure.

Instead of accepting a liability, a difficulty, and working around that blemish, each person thought of the goal first, where he or she was headed. MacCready wanted a human-powered plane that could fly. Edison wanted a long-burning light. And NASA searched for the most effective way to create a building that could sustain itself on this extraordinarily generous planet.

What every solution required was an entrepreneurial spirit. During a trip to China, Bill heard an executive from Microsoft tell the audience of an official joint China-U.S. innovation summit about how Microsoft studies employees and potential employees to assess entrepreneurial instincts. When interviewed, 50 percent of the potential employees said they were entrepreneurial, that they would take risks. In actuality, they found only 1 percent showed willingness to take risks once engaged in an enterprise. The rest were risk averse. They wanted to make sure they got their paycheck every week, and that was their focus.

The 1 percent of interviewees who actually were entrepreneurial disregarded risk. They put everything into what they were doing. When they failed—and of course they did—they just got up and tried again. They were relentless in their pursuit of their passion and their dream. Edison and MacCready were by definition entrepreneurial inventors. Edison is said to have tried thousands of substances for his filament. When someone asked him why he kept trying after so many failures, he apparently said something to the effect of "I haven't failed. I have found so many ways that don't work.

Why would I stop now?" MacCready purposefully designed his plane, even in its trial stages, to be taken apart rapidly. The reason: so he could fail and try again hundreds of times without wasting time. He sometimes failed several times a day with different versions of the same craft, all quickly reassembled.

Since the publication of *Cradle to Cradle* and presenting our strategy as a platform for continuous improvement, we have had the opportunity to work with hundreds of companies and organizations, entrepreneurial and conventional. We have seen several major problems—or misconceptions—that have hampered implementation opportunities of Cradle to Cradle design. We want to talk about one now: the role of values and metrics (by metrics we mean measurements and benchmarks) in design.

Dream What You Want and Let the Numbers Follow

Often businesses and government agencies will put metrics first and let all the other parts of their plan follow. Let's take, for example, building housing for people in Haiti, victims of the earthquake, now being relocated to a region of possible employment. The metric directing that effort is likely to be to house the most people at the lowest cost.

The tactic for lowering that cost customarily would be to diminish the amount of materials used—more wood or cement or steel or composite board creates more cost. The strategy for diminishing the amount of required materials would be to compress all dimensions to the absolute minimum—a ten-foot-tall interior wall is more expensive than an eight-foot wall, for example, so lower the ceilings. The goal would be to create the smallest structures that are still habitable. So this is how the usual big-picture strategy looks before it is upcycled:

1.   Metrics of efficiency: House the most people at the lowest cost.
2.   Tactics: Reduce the amount and cost of material needed for construction.
3.   Strategies: Compress dimensions to save square inches and feet of materials.
4.   Goals: Create the smallest habitable structure for a human being.
5.   Principles: Apparently missing—it's too late to be talking about principles when you have already developed your metrics and your tactics and your goals.
6.   Human values: Unexpressed.

Look at 5 and 6. Where are the principles of the operation of the system, and the values expressing themselves within this system?

In the case of creating housing in Haiti, one can suppose that the larger societal value probably was to provide shelter for people who didn't have shelter. But who knows? The metric target became so all-encompassing, and tactics and strategy shifted so constantly to try to fit into the metric, that there was no time or headspace to consider the value or even consider how "shelter" is defined: Is it merely a roof over one's head? Or is it a safe and pleasing place where a human can thrive?

All sorts of distractions got in the way: The cost of composite board or cement was fluctuating, deadlines were pressing . . .

In the short term, a businessperson or executive's guiding principle is almost never values. Rather, it's benchmarks against existing conditions—15 percent off, $50 million saved. A financially successful quarter.

But long-term solutions cannot come merely from measurements such as these. They will come from innovation. And innovation, by definition, cannot be benchmarked. It's not merely an improvement on a flawed ("less bad") system. Wikipedia and Google probably didn't benchmark the *Encyclopedia Britannica* and say, "We will have 200 times as many entries." They reinvented the whole notion of an information resource and access.

So let's start again: Put values right up there at the beginning of the decision-making process.

Interestingly, by clearly identifying and stating your values, you can drive an innovation process, which falls then into the steps of setting principles, goals, strategies, and tactics. This doesn't mean that you can't begin where you are now, today. Certainly, redesign is preferred; however, "values" can be placed foremost, and existing process or product can be modified piece by piece until total design can reoccur.[4]

Here is a different plan to begin:

1. Establish the value or values for your company's engagement with the world.
2. Then work with teams to establish your principles.
3. Then develop goals to realize those values.
4. Then develop strategies to meet those goals.
5. Then develop tactics to execute those strategies.
6. And, finally, develop metrics to measure the effectiveness of those tactics.

If we looked at this through the lens of Bill's NASA space station, it would appear as something like this:

1. Values: To design an environment to support both human and planetary well-being—with safety, excellence, teamwork, and integrity.

We even write that into our certification process for goods and services. *We understand the real world* . . . and we're still saying you can get to Cradle to Cradle.

2.   Principles: the Hannover Principles.[5]

3.   Goals: Build a working structure that can sustain terranauts and its operation, all within conventional federal budget and schedule constraints.

4.   Strategies: Blend the building with free natural resources readily available and use Cradle to Cradle and other considered materials.

5.   Tactics: Bring together the best scientists and technicians and humanists to brainstorm innovative solutions. Use the sun for light and power. Use the earth and water for cooling. Use air and its movement for circulation and temperature control.

6.   Metrics of efficiency: We now know that Sustainability Base, still being perfected, has reduced its energy consumption 90 percent and its water use 87 percent compared to a building of similar size—"less bad." But it also has the capacity to purify its own water, to accrue more renewable energy than it needs to operate (up to 120 percent), and to feed that extra energy to the grid—"more good." Imagine a building that uses renewable energy for its operations and, to offset the energy used to make the building, contributes extra energy to the grid!

If NASA had started with a metric marker, it might not have gotten to such large numbers so quickly. First of all, it probably would not have set a goal as high as 90 percent and therefore would have been satisfied with a much less efficient building. It would not have innovated as much because it would have simply worked at making a traditional energy system "less bad."

    And you wouldn't have been able to open the windows.

---

5.   See our Hannover Principles on page 10.

It sounds funny, but it's true. If their only goal had been to use, say, 30 percent less fossil fuel, those windows would likely have been sealed shut. But because NASA's value was to support human as well as planetary well-being, they encouraged a design that allows employees to hear songbirds, if they wish. And the project still came in with those metrics, on time, and at budget.

If design is a signal of intention, even small steps in the right direction, with a positive and anticipatory framework, can create a ripple of effects.

For a company, a declaration of intention, loud and clear, can often be influential enough to help drive innovation. It costs the executive nothing, yet it tells people (staff and employees as well as customers and community) where the company wants to be. When a leader charts a course for the future, stating the value can jump-start enthusiasm, activity, and perhaps even the policy change required for that future to become reality sooner. It is helpful to articulate the good (e.g., by 2020, our goal is to power with renewables) rather than just articulating the bad (e.g., we will reduce our use of fossil fuels).

The CEO can appoint an anticipatory design team to begin this process.

The CEO can make sure that the team is cataloging all the opportunities for renewable power—tracking the costs of new technologies, the price of electricity, and the factors specific to running the business (e.g., calculating how many roofs of their multiple facilities are in sunshine, how many parking lots they have, what kind of real estate they own, etc.); then tracking the projected cost to deploy these new technologies, what the utility will pay for excess energy generated to be sold up to the grid, and what the tax benefits are. Perhaps the team in charge bids out the job every three months, watching for price fluctuations, so that everyone sees how serious the company is about this. *(Is the cost of solar energy cost-effective yet? No? Okay, we'll keep working on it.)*

The truly forward-thinking company would create an algorithm (like a Google alert) that notifies the CEO the moment cost-effective, renewable energy is available to use on-site or as locally as possible. The instant that happens, the company can be ready, willing, and able practically to push a button and switch over.

In some cases, the bottom line is more of a top line. When we advise Silicon Valley companies trying to recruit the best engineers from around the world, we recommend that they look at their renewables investment as part of their recruiting budget.

Consider this: If a solar-powered energy system for the company costs half a million dollars more than a conventional fossil-fuel-based system in the short term, and earns out in seven and a half years, a company might question the investment, even if the money will eventually be returned in seven and a half years. Who wants to pay half a million dollars more for something, no matter how much revenue the company generates overall?

But in Silicon Valley, one great engineer can generate $1 million a year in revenue for a company (really). Therefore, if even just one of the candidates the company is trying to recruit says, "Yes, I have chosen to work here because I want to be able to tell my kids that I work at a company fully powered by renewables," or, "Yes, I want to be part of a forward-thinking company," then that investment in renewables pays off in the short term. The company won the recruiting game and the new engineer is worth a great deal of revenue, right away. The company will have met its bottom line and then some.

Sometimes the company's values are simply valuable.

Metrics You Never Dreamed You Could Improve

In *Cradle to Cradle*, we talked about the fabric that we designed for Designtex, a division of the Michigan-based company Steelcase.

When asked to design an upholstery textile, Bill wanted a product that would express his core value of how to love all of the children, of all species, for all time, to create a *healthful* fabric. He suggested the team could use the Hannover Principles for design guidance and could use Michael's chemistry to achieve the goal.

This idea of making healthful fabric was not an obviously achievable goal, given that a good percentage of conventional fabrics contain chemicals undefined in terms of ecological and human health. Trimmings and loom clippings often have to be disposed of as hazardous waste. Fabric dyes often contain toxins, even such heavy metals as cobalt or zirconium.

The usual way to reduce toxic load is to filter out dangerous substances—mutagens, carcinogens, endocrine disruptors, persistent toxins, and bioaccumulative substances—at the end of the process. We took a different approach, eliminating these dangerous substances at the beginning. It's just easier, more efficient. In essence, we were designing "upstream," where it is much easier to control the process than "downstream." As Bill said at the time, "Let's put the filters in our heads and not at the ends of pipes."

Working with the mill, Rohner, the project team chose a mixture of safe, pesticide-free plant and animal fibers—wool and ramie. With the help of a European chemical company, we eliminated from consideration approximately 8,000 chemicals commonly introduced in the textile industry, which in turn eliminated the need for additives and corrective processes. But that wasn't good enough: We went even further and chose 38 *positive* ingredients from which to make the entire fabric line.

Now we come to metrics, and this is where the abundance of upcycling comes into play all of a sudden. The water coming out of the factory was as clean as the water coming into the factory—in fact, clean enough to drink.

As we described in *Cradle to Cradle*, because chemicals without toxic characteristics were involved in production, regulatory paperwork was no longer necessary or required. Employees who had worn gloves and masks for a measure of flimsy protection against workplace toxins could take them off. Space previously used for the storage of hazardous chemicals was now available as additional workspace and even for recreation. The fabric trimmings could be used to safely nurture the soil, as mulch for the local garden club, instead of being shipped to Spain as "hazardous waste."

If the company had simply worked on reducing the amount of toxins in their fabric, none of these effects would have been achieved (you'd still have regulations, toxins in water, masks on the workers, hazardous waste storage and removal, and so on).

The benefits of designing for beneficial emissions of clean water (and clean soil and clean air) from the outset were profound. The product was of higher quality and, importantly, cheaper to make (aesthetic and economic benefits); there was no need to get overly caught up in complex environmental regulations (economic benefit); mill workers stopped wearing gloves and masks to protect themselves against workplace toxins (ecological and social equity benefits—fairness); and the water was clean and drinkable.

Because values had been put at the start of the design process, based on the Cradle to Cradle vision of biological and technical cycles, the product and, indeed, the workplace and ecology around it, had been transformed. The fabric has won awards . . . and garnered customers, who feel good about purchasing a beautiful thing that does not harm the world but instead has been designed *for* it.

In the years since that project, Steelcase has not lost its zeal to make everything it produces, and the way it does so, an embodiment of its values. In 2003, it gathered a team with the goal of developing a chair. This chair would, of course, have to be beautiful, and it would not just be comfortable but would adapt to the way you sit, reducing fatigue and discomfort. It would be original, even iconic. It would also need to be completely environmentally sound and very profitable. The values of the company were set from the start. The Think chair would embody every one of its design principles.

At many junctures during development, a material was chosen for its performance but turned out to be a substance rated "high risk" on McDonough Braungart Design Chemistry's and EPEA's material health scales—what we call a "red."

In those cases, the product leader would suggest replacing it with a "yellow" or "green" (shorthand for substances of, respectively, "low to moderate" or "little or no" risk).

It wasn't always easy to find a suitable substitute whose performance was comparable. And here we come upon one of the biggest hindrances to redesign—in fact, for the most part, the reason industry has long continued to use hazardous materials. It is not out of some dastardly indifference to the environment or worker health. Designers choose these materials or substances because they perform so well, with few (and sometimes no) nonhazardous, cost-comparable, performance-comparable substitutes that the companies know about. Finding a substitute takes work and time.

Once that substitute is found, however, the advantages, benefits, and gains are worth it.

With the Think chair, some difficult choices had to be made, even with the core values decided upon. What happens, for example, when the choice one must make is between two environmental concerns?

The Think chair had been scrupulously measured for its carbon footprint in materials, production, transport, and conventional "life cycle" assessment. Steelcase wanted to keep those numbers as low as possible due to their value of ecological health and safety. Yet at one point, during life-cycle assessment comparisons, the Think chair's energy footprint shot up compared to previous measurements. Looking at the new number, the product leader wondered what had happened. Why did the chair just get so much worse?

Here's precisely what happened: During the research-and-development phase, PVC (polyvinyl chloride), known for its great flexibility but also known to have a carcinogenic precursor and to off-gas toxins when burned, had been replaced with TPU (thermoplastic polyurethane), a nontoxic substance of equal performance. PVC, however, has much lower "embodied energy"—a measurement of the amount of energy needed to make, deliver, and dispose of the substance or product—than TPU. A dilemma: Use the toxic substance that requires less energy or the nontoxic one that requires much more energy? By swapping out PVC, they had solved the toxicity problem; by swapping out TPU, they had solved the embodied-energy problem.

Steelcase, ever sensitive to ecological questions, decided from that day forward that their value would never be to choose a toxic material because it had a better energy footprint (it spent less energy in its creation). They went with the TPU. This makes sense, since it is much easier to employ renewable energy in creating a product (as you will see in the next chapter) than it is to detoxify what has been toxified. In some ways, you might say the metric *was* the value.

The Think chair, which came out in 2005, won numerous awards for its elegant, innovative design. It was (and is) available in many styles, fabrics, and colors; is lightweight; boasts a stellar environmental profile; and contains no PVC or HCFCs (hydrochlorofluorocarbons). A minimal amount of glue is used in its assembly, and it can be disassembled in five minutes with common hand tools for safe and easy repair, reconfiguration, return,

McDonough Braungart Design Chemistry: How We Help Organizations Optimize

We have been motivated to create abundance at all scales. In the realm of creating abundance at the molecular and product scales, we founded McDonough Braungart Design Chemistry (MBDC) in order to work with companies that wanted to change the way they work and make things. Through MBDC, we developed the Cradle to Cradle values and principles that have been used to spur innovation, differentiate businesses, and benefit people and planet. We help organizations develop a Cradle to Cradle road map and strategy that go beyond minimizing harm and move toward a wholly beneficial impact on economy, ecology, and equity. We provide services to inventory, assess, and optimize products and processes and offer support to achieve product certification and communicate achievements. In 2010, working closely with EPEA in Hamburg, we put the certification program into the public realm by founding the Cradle to Cradle Products Innovation Institute (see pages 198–99 for more information about its mission and purpose).

We believe these companies that start the process by entering certification truly deserve to be honored, even if they are only at the beginning of the process. They dared to lead the way. They dared to innovate. And their solutions will allow for a cascade of cleaner designs across industries.

Because of the complexity and demands of the certification process for a company, we mark its intention—or, rather, its intention joined to its effort. It is not reasonable or realistic to expect a manufacturer to jump to fully clean production overnight. A company can however enter a process that helps move it from cradle-to-grave production to Cradle to Cradle production. The company starts at the beginning. Over time, it works its way toward endless resourcefulness.

The certification protocol we created helps signal to consumers that the manufacturer has begun the journey. We also want certification to be a badge of honor for the manufacturer who takes on the upcycle goal.

For that reason, we originally named the levels Basic, Silver, Gold, and Platinum (the Bronze level was recently added for Version 3 of the certification criteria)—and all of the other levels need to be achieved to move up to the top designation of Platinum.

You may find it ironic that we named these levels after heavy metals. Platinum can cause respiratory allergies; soluble compounds of gold can be toxic to the liver and kidneys. If introduced into the biosphere, these metals might not be entirely welcome. But we wanted our clients to be able to display an honor that most people understand. Metal levels are commonly used to indicate rarity: This individual or this company had achieved a state of excellence thoroughly uncommon in our society—as rare, say, as platinum. Perhaps in the future, the designations will indicate rarity of a different kind: Basic could be red wolf, Silver might be the Vancouver Island marmot, Gold the Amur leopard, and the ultimate one could be the apex of the Yangtze River dolphin (although that version of Platinum, sadly, might have to be revised in the future, since fewer than 14 remain . . .).

To begin the certification process, we help a manufacturer understand what is in its product to start with. Oftentimes, a manufacturer knows the basic materials of a product, what kind of plastic or metals are involved. However, it often doesn't know all the important steps its suppliers and subsuppliers take in making the product and which chemicals are introduced with the addition of dyes, fasteners, metals, plastics, or finishes.

At the Basic level (under Version 2 of the certification program), manufacturers are just engaging in this process and only beginning to understand the complexity of their products and the potential impacts of the materials and manufacturing processes. At least 95 percent of the product must be identified down to the parts-per-million level and evaluated for safety to human and environmental health. For an office task chair, this process alone could take five to six months and involve hundreds of suppliers, many of which have proprietary formulations that have never before been evaluated by a third party. There can be no PVC, chloroprene (the chemical name for neoprene), or any related chemical in the product at any level. The product must also be designed for a biological or technical cycle. At this level we say the product is "defined" and we understand the starting point for the product's Cradle to Cradle journey. The company must develop a strategy to optimize any problematic ingredients and must show improvement within two years to maintain certification.

We like to say that the criterion for the chemical filter is that the chemicals used will not accrue in mother's milk. This essential criterion is what has formed the foundation for our expanded Banned List, developed for the latest version (Version 3, being used now) of the certification program. We like that concept because it gets down to the basic idea of whether or not the product is toxifying the nutrients necessary for new life on this planet.

Conventionally, companies might prefer to be given a list of forbidden chemicals. Sometimes that desire for a specific list is simply a way to figure out how to make a slight change without answering the core concern; for example, a chemical can be tweaked by one or two atoms so that it no longer qualifies for the list, but it may still have the same toxic characteristics as those substances on the list.

But that's not truly a sufficient response. With chemicals and elements, it is where the material is used and how it is used that determine whether the material is problematic. Our assessments look at the direction and the intention of the material, as well as the material itself.

When we sit down to consider such specifics, the conversation inevitably turns to larger issues.

To date, much environmental activism has been devoted to pointing out chemical hazards—vital work, of course. Yet few organizations have taken the next evolutionary step, to go from looking at what is bad to using what is good.

What can we use? What would be beneficial?

We have always wanted to identify a list of positive chemicals that can be employed in creating a multitude of products, not just indicate negative elements and catalog chemical problems. Historically, the EPA has called its Acutely Toxic Chemicals list its P List. But we wanted to take a more optimistic approach. Our P List is our Positive list.

For example, if we are working on a textile, we scrutinize every material down to the fibers, dyes, and auxiliaries. Then we create a list of substitutes that we know do no harm, and might even be beneficial. We have had remarkable experiences working with companies on this process. Small discussions have tended to create cascading beneficial effects in the company and in the world. For example, we started working with Herman Miller, then Steelcase, two of the largest U.S. furniture manufacturers; next thing we knew, we were working with Allsteel, Haworth, and others as well. We were able to transform the industry by working with many common suppliers, encouraging them to innovate and change what they make to meet the demands of these furniture makers. Byrne Electrical, for example, innovated a PVC-free wire-jacketing product for Steelcase that is now available for other manufacturers to use. They only invented it because of the demand created by this program.

Recently, we were talking with a client about a small bit of metal used in the dispenser/plunger of her bottle. Because it was not designed for disassembly, it could not be reclaimed easily. What could be done about it? The metal was not irreplaceable, as it turned out, and we were able to come up with alternatives.

More Important, the conversation bloomed into one about packaging and recyclability. Was there a way to redesign the dispenser mechanism so that it was fully reusable? Could the packaging be made fully reusable? If so, what economic opportunities were created by having fully reusable packaging?

The dialogue eventually morphed into a still larger one—about perspective on design and decision making. These are the sorts of conversations that can be so fruitful—the result of upcycling one piece of the whole design picture.

reuse, or recycling. On its website, the company tells you exactly which items go into the standard town recycling center and where you can bring or send other parts for more specialized recycling. It became the first product to receive Cradle to Cradle certification.

And it has made the company lots of money. Indeed, the Think chair has been so successful that the company has made it its goal to have every one of its products, upward of 70, earn Cradle to Cradle certification. More important, everything it makes will embody Steelcase's values.

Would any, much less all, of these numerous achievements have been possible as afterthoughts? Absolutely not. But once they had begun with such values, the additional profit of unintended positive effects (rather than unintended negative effects) cascaded.

Rather than starting the way business usually does, with metrics—the bottom line—and losing steam or focus before ever getting to the most important point, values, this plan starts with values and makes its way to metrics. The later you consider values in this process, the less likely values will be considered at all. Metrics, unlike values, are in no danger of being ignored because there will always be some bottom-line measurement of whether one is achieving what one set out to do.

To look at it in reverse, this plan sets the context for making informed short-term decisions because they are based on inviolable values and principles.

These goals actually prompt innovation all the way down the line. We have found that if we work with manufacturers and talk about values first, we can produce far higher levels of innovation and performance than if we benchmark. For example, the caster wheels on an office chair like the Think chair are traditionally made with leaded steel—meaning that lead is added to make the metal easier to form into that particular shape. The reason a manufacturer likes the more pliable metal is that it speeds up production; you can get the part quicker.

So we went to the factory that makes those wheels and asked if there was any way at all that it could make a caster out of positively defined materials. Was there any brilliant idea it might have for how such a dramatic, healthy innovation could be possible?

And you know what they said?

"No problem."

It turned out that no one had ever asked them to make such a thing and all they needed to do was make simple steel casters—take out the lead. The secret to getting a healthier addition to this chair was simply to ask. If Steelcase had not started with values at the top of their six-point plan, no one might have asked.

The possibilities here are very exciting: Our world can be made truly clean, safe, and healthy when designers, engineers, and businesses embrace innovation that grows the good, not by continuing conventional production, making things somewhat "less bad" and watching the metrics improve bit by bit until you reach what popular eco-efficiency enthusiasts oddly and perhaps even bizarrely call "mountain of zero." Instead of this confusing perspective, what if ambitions were stated as: "How can I select and use 100 percent positively defined materials and renewable energy? How can I increase prosperity, celebrate my community, and enhance the health of all species? How can I build a space station for the earth? How can I make my yard a habitat? How can I design a chair of safe materials that can be endlessly reused?" The results can be astoundingly positive and enriching.

The Butterfly: What's Next Might Be Hidden in Plain Sight

We want to let you in on a secret. Okay, it's not really a secret; it's more the sort of thing Hildegard of Bingen would have liked, and it relates to these ideas we have been discussing: innovation and values. It involves looking around, seeing the greening of the world, and thinking.

When we were examining dyes used in polymers for carpet fibers and fabrics, we kept stumbling upon a problem: We admired many of the colors available, but they contained heavy metals. Fabric makers have been challenged to find dyes that produce the gorgeous colors necessary for aesthetic appeal without adding a contaminant. And we definitely did not want to compromise the intensity of hue.

But, we thought, the natural world radiates with vivid colors. How do birds do it? Their wings don't contain heavy metals. How can we emulate the rich jewel tones of a peacock or a scarlet macaw or a monarch butterfly? Emulating the designs of nature (called "biomimicry," to use the term popularized by Janine Benyus) can be incredibly fruitful, and designers are doing this not just for aesthetics but also for performance.

So we studied bird and butterfly colors. What we found was that the vivid color of bird feathers is produced by light. Traditionally, working with textiles has meant making a subtractive color: adding pigmentation and then working toward black with color reflection. But the colors of animal fur, hair, and feathers are created by light refraction. Simply put, exotic birds and blue jays are flying crystals: Their feathers are nearly colorless. If you look at a feather of one of these birds under a microscope, you'll see that it is transparent. It refracts the light to achieve distinct colors, the way a prism does. Butterflies work this way too; they are shingled to shimmer like crystals.

So who will develop a new set of polymers that are refractive, so the color will actually be in the reflected light and not in the dye? How about a clothing designer who makes clothes that are unpigmented but that angle pure light so that the colors are as breathtaking as any produced by a chemical dye? The scarlet macaw feathers will clearly fade back into the soil, as will iridescent butterfly wings and the colorful flowers in the meadow.

Now that would be an innovative answer to the question "What's next?" It might be looking more curiously at what is already flying by.

Wind
Equals
Food

Chapter 3

We have talked about carbon, the building block of life, and of the desire to keep carbon where we most need it—earthbound. We have discussed how the upcycle asks us to conceive systems that maintain carbon in loops of endless resourcefulness. We would like to discuss now the *how* of keeping carbon on earth. Let's touch down for a moment in Scandinavia, in a place and at a time where the photosynthetic act of carbon dioxide being converted into carbohydrate is sometimes difficult to achieve.

The problem, of course, is that in Denmark, during the winter, it can get dark early. At least that's how it seems sometimes to Danes when December rolls around and the days shorten to a scant seven hours. It's amazing how much one can start hungering for a breath of spring, or the first greening of a plant, or a fresh strawberry. Danish supermarkets are, of course, full of options due to the wonders of international markets, but we considered what we could do to promote happiness of the revivifying kind for people where they live and work every day.

Over the years, we have been asked several times to consult on the concept of Cradle to Cradle islands because, with their ocean boundaries, islands are contained systems with clearly identifiable inputs and outputs. Michael helped conceive systems for an EU-funded project for the islands of the North Sea. We also worked with a small island off Denmark. There, along with designers and engineers from six different countries, we were asked to devise concepts for a science park. We wondered what it would mean to help people get a chance to feel as if they're a part of rejuvenation of the planet anytime, any month.

In the days of plentiful daylight, the science center lobby could benefit from shades. But Bill, an architect, thought, *Wouldn't it be marvelous if the shades were plants growing behind glass, thin horizontal hydroponic planters?*

Then he thought about winter.

The days shorten dramatically beginning in September, so how would these plants keep flourishing? Or how would we get fresh local food in winter?

The answer: another form of solar energy. The wind.

For Bill, this was a conceptual leap, because while it came to him in the context of architectural design, it was the seed of a much bigger idea: What if a darkened island surrounded by water could use wind to feed itself with fresh green food in the winter? When Bill first said that wind equals food, people scratched their heads and said that wind makes it hard for things to grow.

But Bill had worked on wind power in the early 1980s, helping to develop wind turbines that used electric generators to control and break the speed and the angle of the blades instead of mechanical gears. Wind can be used to generate electrical power. Electrical power can be used to generate light. Bill had become interested in LEDs (light-emitting diodes) for growing food and found a botanist who was conducting experiments showing that some fruits and vegetables, such as strawberries, don't actually need the full spectrum of light granted by the sun to do their growing. Strawberries, for example, really are interested in only a portion of the spectrum: reds and blues. We now know that LEDs can be tuned to emit only these frequencies, which reduces the amount of energy needed to power the lights.

So for a portion of the energy needed to create the full spectrum, light can be emitted that strawberries love, which makes them grow big as a fist and sweet as honey. This can all happen at night when people are home in bed. The strawberries don't care what time of day they get their dose of light. They will happily receive a supply at midnight, when a gust of wind comes through. And in the morning, when people get active again in the building, the plants are flourishing. You might get to have a fresh strawberry at the morning meeting.

We're sharing that story to underscore a larger point: We think discussions about energy can expand into new dimensions instead of being laterally focused on conventional practice. Modern society and most designers look at energy as a separate phenomenon unconnected to other functions, an isolated need rather than a means to an end. Because of that conceptual limitation, short-sighted decisions cause missed opportunities for upcycling.

If you think about electrical energy from its power source, through its grid, to its potential multiple-use cycles, and on to the ultimate needs for power in the first place—warming your house, heating water, lighting your desk lamp so you can do your homework—you can begin thinking of how to optimize the system for more abundance, more productive use. This kind of full-circuit thinking can expand your inventiveness.

This is upcycling energy.

At this moment in history, society has begun to reckon with energy shortages and with the desperate need to wean itself off fossil fuels—to use less. But we want to restart the discussion with a use of energy for pure pleasure, because such an idea is possible with an effective renewable source. No one *needs* strawberries at his morning meeting. No one *needs* to see green plants lining her windows year-round. But they sure do bring a smile to people's faces. If those LEDs were powered by fossil fuels, everyone might declare the strawberry project a waste. But if free wind input, economically, ecologically, and ethically delivered, can make a continuous source of pleasure through the year, then who would gripe with a ripe local strawberry in winter near the Arctic? [6]

---

6.   It is worth mentioning that the Icelanders have long been proud of being able to use geothermal energy to heat and light greenhouses that can grow tomatoes and even bananas near 66° north latitude.

These energy concerns are not new to us. In *Cradle to Cradle*, we made a point of focusing on products and material flows because so few people were considering materials and because so many people were already debating energy efficiency. Then, as even today, it seems most environmentalists were focused in that direction.

It wasn't until we developed the Cradle to Cradle Product Certification program over the past few years that we started publicly articulating our energy position and integrating it into our design protocols. Clearly, energy is a key question in any work.

When many people who are focused on energy read *Cradle to Cradle*, they said, "This is all well and fine about redesigning products, but the glaring issue is power. Without a Cradle to Cradle fuel source, the end product can only be an improvement of an ineffective system."

Basically, they were highlighting something that even Edison, back in the 1800s, understood. While he searched for a workable bulb design, he was thinking too about electric generators and the grid. He understood the lightbulb to be only one outlet in the larger issue of how power could flow to homes and businesses and focused his efforts on direct current (DC). Nikola Tesla understood this issue too and invented the alternating current (AC) generator, allowing longer transmission distances.

So we take up the question of energy and its distribution now because it is one that human society needs to engage. The solutions will require us all. And no product can be considered truly exquisite or well designed unless the amount or type of energy used in its production has been considered.

As society goes forward in devising solutions, we believe it is key to bear in mind how important it is to respect and understand the end goals—the real needs—of all the players; that the energy is extracted in the cleanest manner possible; and that we use energy that is ever replenishing.

And, finally, to expand our thinking as we consider solutions, let's reflect on the different ways we typically think about energy and about how it serves us. Sometimes energy might conjure the image of a gallon of gas that powers our car for a certain distance. Sometimes we think of a kilowatt-hour that lights a bulb. Other times—and here this exercise begins to get curious and opens up creative opportunities—we might look at energy as calories, which measure heat potential. We even talk about calories for human consumption. In the remainder of this chapter, we will look at energy from a richly diverse set of perspectives and see where the modes of energy might meld and foster growth, foster the upcycle.

Complex Systems: How to Cook Naan Without Killing a Tiger

Let's look first at the end goals of energy. Energy concerns are complex, with many competing stakeholders. It does us all a disservice to deny the merit of their needs. But we believe that fully understanding those needs might allow opportunities for innovation. A person turning on a lamp is not asking for 100 watts of energy; the person is asking for beautiful light by which to read a book. If one can address that core need in a delightful way, the person is happy.

As an example of how to broaden our thinking, let's visit India and hear a story, one with sister stories in places such as Nepal, that demonstrates how one can deal with complex competing needs. In a village nearly six hours south of New Delhi, a conflict arose between humanity and ecology. Local farmers were going into the Ranthambore National Park in Rajasthan to cut down trees for firewood. They were also sending their cows into the forest to graze.

The national park served as a rare tiger habitat. As the villagers cut down the trees, the forest receded. And the tigers now found themselves in proximity to herds of grazing cows, which, of course, they attacked for food. To save the cows, the farmers looked the other way when poachers came to kill the tigers for their pelts.

The farmers wanted fuel for themselves and food for their cows. Conservationists wanted to preserve the forest and the tigers. What an unfortunate sequence of events—and how interconnected they all were. How could one solve a problem with such complex effects?

Park steward Fateh Singh Rathore's first solution was to put up fences. But the villagers simply climbed over or broke them. (Regulation as temporary fix, not solution.)

The solution had to be as interconnected as the conflicting needs of the environmentalists and villagers and tigers and cows, and it was brilliant, bringing what appeared to be separate agendas into a productive community of profit for all: for the individuals, the ecological system, and the species.

The steward's son, Dr. Govardhan Singh Rathore, who inherited oversight of the park, first set up a health clinic to improve overall conditions locally and build goodwill. Then he helped the villagers breed cows that produced more milk with less feed (less vegetation needed). Then he suggested to the farmers that, rather than cut down the forest for fuel, they gather cow manure and transform it into fuel and fertilizer using biogas plants. The farmers could keep the cows closer to home (and wouldn't have to venture into a forest with ferocious tigers). They could stop cutting down trees.

At first the biogas plants were provided free to the villagers, but Rathore soon realized that ownership conferred responsibility. Now villagers buy the biogas plants for approximately 3,000 rupees (about $54), and 600 of them operate throughout the area. Rathore also planted new trees to replace the ones cut down and paid villagers if they could keep their assigned tree alive; the pay rises incrementally for each year the tree survives. He offered the poachers free education for their children if they stopped poaching and gave them camels to create income from milk and to use for transportation. Now the poachers help protect the tiger habitat; it is in their best interest. We see almost exactly the same situation with converted big game poachers in Kenya.

Let's look at what happened here: Energy was the needed resource. The farmers valued fuel and cows. The preservationists valued the forest and tigers. Both groups valued success. And Rathore, the one who conceived all this, valued all of these elements—people and community, nature, a fruitful and enjoyable life for all involved.

In this instance, what appeared to be a conflict between a small rural community's modest "economy" and the ecosystem was resolved by a creative solution.

Energy economies large and small, of the most modest means and of the greatest monetary value or assets, can be upcycled. First, clearly identify the triple top-line value for each constituency, then aim to protect those values while reconfiguring the system. The farmers in Rathore's village wanted energy for cooking. They weren't particularly wed to the requirement that the fuel come from the forest. The preservationists wanted the survival of the tigers, but they didn't want the community around the forest to perish. By understanding individual and mutual values, and attempting to address them in new ways, we may discover unexpected solutions for energy challenges.

Getting the Energy You Need . . . Cleanly

Another important aspect to consider in energy solutions is *how* we glean energy. We want you to consider pigs for a moment.

In the United Kingdom and Europe, farmers traditionally sent their pigs into forests to forage. This custom served many functions: Pigs could eat a wide variety of foodstuffs, the farmers didn't have to pay for extra scraps to feed them, and the pigs' foraging helped control weeds and small pests and recycled nutrients for improved soil quality. In essence, pigs were calorie and nutrition skimmers. The pig went into the forest—the perhaps frightening, brambled forest—and came out with the best pickings of that forest in its belly. When the farmer killed and cooked the pig, he and his family benefited from its expert foraging. The pig was an efficient machine for capturing the riches of the forest for human consumption, for gathering energy for human consumption.

The only problem came when the forests became seriously depleted and authorities had to come up with ways to keep the farmers and their pigs out.

We can consider our energy needs in this light. How can we reach the best resources, no matter where we are? How can we collect the energy we need without literally and figuratively depleting the forest?

Capital vs. Currency: Spending Resources

Finally, before we get into specific energy discussions, we think it's important to consider which kind of energy we might prefer. An appropriate way to differentiate between what we consider the better sources of energy comes to us from the world of economics and frequently crops up in our work advising companies. One has to define what in one's world is currency and what is capital and how goods move between these categories.

The Peruvian economist Hernando de Soto in *The Mystery of Capital: Why Capitalism Triumphs in the West and Fails Everywhere Else* dramatically describes the dimension of informal economies in which the poor face great difficulty because of their inability to accumulate capital. Capital does not "flow" but is stored, saved, invested, or embodied for future deployment. As de Soto says, "Capital is not the accumulated stock of assets but the *potential* it holds to deploy new production" that "must be processed and fixed into a tangible form before we can release it." Currency lubricates the wheels of commerce and is both a lubricant and a measure of flow.

De Soto's point is extremely relevant to the upcycle, to how humans handle planetary materials and manage nutrients. People are currently burning, burying, and otherwise dispersing and contaminating their earthly capital. We're treating capital like currency. Here is the difference: Say you have a goat. You get hungry and you eat it. The goat is currency. You have converted it to human food (caloric energy, by the way) and it is no longer with you. You have nothing left.

What if you have an apple? It's crisp. It's here. You eat it. Again, it is food. Let's hope you did something good with those calories because you don't have another apple. It's gone.

But if you have a herd of goats or an apple orchard, you have capital. These can go on perhaps forever with endless resourcefulness, multiplying, bearing fruit, providing you with more energy and abundance for long centuries.

Since the oil crisis in the early 1970s many commentators have made a devastatingly simple point: When humans use petroleum, they're using ancient sunlight. In the previous chapter, we described how fossil fuels represent ancient biosphere—plant and animal life from long ago.

How was that biosphere created?

Photosynthesis.

What lies beneath earth's crust is, by extension, stored-up ancient sunlight.

So we can ask again and again: Why aren't we using current sunlight instead? The sun's energy—solar income—is recurring; it's the only income (other than meteors and cosmic dust) that our planet receives. Fossil fuels, on the other hand, are limited in quantity and must be dredged up, with laborious and deleterious efforts. In other words, fossil fuels are actually capital, and we could be using currency.

For decades, humans have dwindled the supply of oil when another energy source has been there all the time, good enough for nature to use for most of its processes. Now that photovoltaic panels and wind turbines have become cost-effective commodities, they can be considered part of our real estate, or capital, that creates currency, whereas burning fossil fuels is capital used as short-term currency without accounting for liabilities—its deleterious effects to the climate and human and ecological health.

As we discussed, it is not wise to put valuable carbon into our atmosphere and our oceans, where it serves no positive use and does real damage. It is also not wise to use and rely on fossil fuels as if they were currency, when they represent a finite material on the planet. By using the planet's fossil fuel "life savings," so to speak, to meet daily energy needs, societies become entrenched more deeply in a system that can't perpetuate itself. There is no good reason to squander this capital when humans have so many energy resources that are capable of rejuvenation.

Fossil fuels optimally would be the nest egg—used only in emergencies, for example, to create the important medicines that require benzene, such as acetaminophen, antispasmodics, antibiotics, and such. It's probably best not to use benzene to drive an inefficient car.

The Ultimate Nuclear Reactor

If human beings don't use fossil fuels, what's the best solution? We are often asked point-blank about more common and controversial ideas for replacing fossil fuels, such as nuclear power. Many environmentalists and technologists have recently come out in favor of nuclear power because they believe it does not involve massive CO_2 emissions in its operation phase.

In the United States, nuclear power has its own regulatory commission, with a budget of more than $1 billion a year. As we have said before, a regulation is an indicator of the need to redesign. Not only that, but regulations herald high financial expenditures, because it costs money to put the systems in place to meet those regulations. Given that, the financial cost of nuclear power is nearly prohibitive. In August 2012, Jeffrey Immelt, the chief executive of General Electric, said that the cost of building nuclear reactors was so high that it had become "really hard" to justify them compared to other energy systems.

Beyond any question of profitability (or lack thereof), we have a more fundamental design concern about nuclear power. We prefer humble systems wherein if their designs show the mark of human error, the result is not catastrophic. We prefer to focus effort and attention on systems that do not burden future generations with remote tyranny. Perhaps in this we are also influenced by the time and places in which we grew up—Bill was five years old when his parents let him see what nearby Hiroshima looked like after the bombing.

So to the question of what we think of nuclear power: We love nuclear power. We are particularly fond of fusion. We think it's a great idea to spend trillions of dollars immediately to access it. And we are so glad we have our fusion-based nuclear reactor 93 million miles away. The even better news—the power is wireless, reliable, and free.

Now, we recognize that turning the nuclear question to talk about solar might come across as glib or annoying to anyone working seriously on the larger energy-deployment models because of the practicalities surrounding current energy sources, fossil fuels in particular. On the other hand, we truly feel this way, and there are plenty of plausible ways to achieve a renewably powered world. It's hard for us to view that goal as unreachable or quixotic when we are working with Walmart, with two million employees, and with Google, with 33,000 employees, to implement renewable energy as their working systems.

Governments like Scotland's have committed to becoming 100 percent renewably powered by 2020. Iceland is entirely locally powered and, as one of its next priorities for optimization, has targeted vehicle fuel with research.

When envisioning what energy could look like in a Cradle to Cradle world, we like to think again about Thomas Edison. In 1931, the inventor reportedly said, "We are like tenant farmers chopping down the fence around our house for fuel when we should be using nature's inexhaustible sources of energy—sun, wind, and tide . . . I'd put my money on the sun and solar energy. What a source of power! I hope we don't have to wait until oil and coal run out before we tackle that."

Not All Renewable Energy Is Equal

We named this chapter "Wind Equals Food" because, in the spirit of the upcycle, we want you to imagine a self-replenishing system of which energy constitutes one part. We didn't call it "Wind Equals Electricity" or "Electricity Equals Food," because those titles wouldn't encompass the full story line. We want you to think about interdependent reuse periods and even interacting loops of materials and energy. In the cycle, renewable energy converts to value, even if the value is simple pleasure—a fresh strawberry to start your day.

Renewable energy is not only an inexhaustible option but also the most delightful option in stimulating design innovation—it is inherently local, is typically silent, and represents a broad spectrum of design options from passive design, such as allowing sunshine and breezes to naturally light and cool a building, to active design, such as using solar photovoltaic cells and wind turbines. In building designs, we have also employed low-grade geothermal. (If we wish to drill, why not drill for renewable energy and not molecules?) High-temperature geothermal is interesting as well (oddly, it is nuclear like the sun—the earth's core harnesses the fission of nuclear isotopes). Iceland, for example, powers 66 percent of the country with such high-temperature geothermal generators.

Of the renewables, solar remains a favorite, in its most basic form and via the other sources to which it contributes—for example, wind (which gets additional help from the rotation of the earth and gravitational forces).

But not all renewables are equal. Just as in the Rathore solution, where the why of each person's action was considered and new routes devised to get there, one must always bear in mind the value behind using renewables—a healthier natural world, more carbon preserved in the soil, preserving underground stores of capital for more worthwhile endeavors.

Sometimes when a renewable is used, the metric of the renewable is considered (a company saying 30 percent of its power comes from renewables, for example) but not the total value. Not considering the value in design causes problems. The ideal solution looks around the corner and understands "What's next?" As Rathore and his tigers will tell you, one person's solution is another's ecological disaster. Or what works renewably here is not renewable there.

As we mentioned earlier, the European Union declared in 2004 that 20 percent of its total energy consumption needed to be supplied by renewable energy sources by 2020. Unfortunately, perhaps, they started with a metric that ultimately may prove to be inadequate to the task. If 20 percent is good, doesn't that imply that more is better and that 100 percent would be ideal? Remember the upcycle chart from the introduction? Why not use it as a guide? Then, when the EU meets its interim goal, the leaders remain aware of the larger ambition—where they are going. They have planned for the whole thing. You need to set your destination.

But the EU, at that moment in 2004, did not have the courageous vision of the explorer. Fair enough. The project of figuring out renewables is hard and different and surprising and uncertain. That project has all the characteristics of exploration and experimentation and innovation. It's difficult. Unfortunately, the EU had driven a stake into the ground before optimizing around the creative opportunities inherent in saying, "We want to go renewable generally."

Part of the problem, as we have been stressing, is that starting with a metric blinds planners to the larger effects of one's actions in a way that a values-based plan does not. Some of the EU strategies did not work so well. For example, one of the ways the EU tried to meet the 20 percent objective was to begin importing palm oil from Indonesia, which they felt could be burned to make a "clean" non-fossil-based biofuel.

Switching to palm oil may result in less CO_2 being released from burning fossil fuels in Europe, which was ostensibly the value of switching to renewables. But Europe's importation of 3.8 million tons of palm oil in one year alone has accelerated deforestation of the Indonesian rain forest after land was cleared to plant palm forests.

Beyond the tremendous loss of flora and fauna, there is, ironically enough, a large increase in the emission of carbon dioxide into the atmosphere. Why? The Indonesian rain forest grew in peat bogs where the wet soil slowed the decomposition of the plant material. To grow the palm trees, foresters had to drain the rain forest land, exposing the decomposing peat, which releases methane and carbon dioxide.

Because of these emissions, Indonesia is now the third-largest emitter of greenhouse gases, behind China and the United

States. Ironically, the lower carbon dioxide emissions in Europe from using biofuel are offset by this other carbon dioxide release in Indonesia.

After the Flood: Creating Jobs That Keep Working

Let's look at another "clean" project: the Kárahnjúkar Hydropower Plant in eastern Iceland. This was essentially an effort by the Icelandic government to create jobs and revenue by harnessing the nation's potential energy production, not for use in powering the country but for Alcoa to make aluminum.

This makes some sense. Iceland knows how to create vast amounts of excess power through renewables. For all intents and purposes, 100 percent of the country's electricity and space heating comes from geothermal and hydro sources.

Iceland knew it could create more renewable power for export. But how could it deliver that tremendous resource overseas? It's not efficient to run cables to Europe to deliver electric power. Essentially, what Iceland wanted was to create a battery capacitor to store the energy and ship it out.

If you think about it, aluminum is similar to a fossil fuel in what it represents, energetically speaking. It is a material used in modern industry that requires vast amounts of energy to create from bauxite. When you see aluminum, the very fact of its existence means that huge amounts of energy were used to bring forth its creation. The aluminum has what is known as high embodied energy.

To create enough power to smelt aluminum, the engineers in Iceland designed a dam that is now the largest in Europe, nearly 633 feet tall, and flooded a 23-square-mile wilderness area to create the necessary vertical drop to power 40,000 average households annually. The Icelandic government anticipated 1,000 permanent jobs from the new smelter, which would be run on "clean" energy (to a population of 309,000, that's a lot of jobs). In 2008, the value of aluminum exports exceeded that of Iceland's fishery for the first time.

But the dam was built in what was the second-largest unspoiled wilderness in Europe (not anymore). Use of the land around the plant is now limited because of emissions from the smelter. Many environmentalists have documented serious damage done by the creation of the dam. Also, huge carbon anodes are required for smelting. This is not carbon-free production.

Now, if we look at the perceived economic value, Iceland's plan makes sense. The government wanted to create 1,000 permanent jobs and was willing to invest $1.3 billion in the hydroplant to make that happen. Alcoa invested $1.1 billion in the smelter. But, given the size of Iceland's capital investment (and the fluctuating price of aluminum), one can ask the question: Is aluminum, fabricated this way—from imported bauxite and with such large-scale hydropower— the most optimized capacitor for the country to be creating?

What if, instead, Iceland decided to obtain aluminum through recycling, which requires 5 percent of the energy needed to extract alumina from bauxite? What if Iceland, using renewable geothermal power, devoted itself to being the aluminum recycler of choice for all of Europe? It has been reported by Alcoa that 66 to 75 percent of all the aluminum ever produced is still in circulation today. How wonderful to imagine a technical nutrient of such endless resourcefulness cycling over and over on renewable geothermal power.

Also Iceland—with its brilliant engineers and scientists— can help lead the renewable economy, as it continues to investigate hydrogen, experiment with high-latitude greenhouses, and create geothermal technology that has a global market. That would be putting Iceland's brainpower and its renewable power into a capacitor to solve the world's energy problems. Iceland would be exporting innovation *and* energy.

Brazil too tried hydropower, resulting in a slew of negative effects. The rain forest wasn't cut down, which sounds good. Instead, the rain forest was flooded to create the requisite water mass and vertical drop to drive the generators.

Unfortunately, wood and other vegetation soaking in the water create methane as they biodegrade. The methane emissions made more greenhouse gases than were offset by using hydropower. Highly corrosive hydrogen sulfides caused by the biodegrading plant matter wreak havoc on the turbines. Those hydrosulfides kill wildlife in the rivers.

Yet the surface facts remained: Trees weren't cut down, which is environmentally beneficial, and money was saved because there was no cutting, which is economically beneficial. But, really, is this good design?

A Renewable Fixing Its Flaws

What makes large-scale hydropower so troublesome is that it does not allow for easy revision.[7] What if the designers and engineers are wrong? How can the structure adapt? If you've flooded square miles of wilderness and built a dam 63 stories tall, it's hard to revise, to amend the system. It is not much different than strip mining or generating nuclear waste in that one has no opportunity to revise. That's

---

7. We are certainly not against hydro overall. Bill, for example, helped restore small-scale hydro-electric plants in Vermont in the early 1980s. We especially prefer "high head hydro," in which essentially the water comes down a mountain in a pipe and spins a Ferris wheel, and certainly we don't mind "run of the river," which doesn't change the course of the stream with disruption or storage. The water doesn't care and the fish don't care. The water is falling regardless. Of course, you can't generate enough power to smelt aluminum from small-scale hydro. We understand that. But it is often difficult to see large-scale hydro as the best option given the other renewable sources becoming available, like distributed solar and wind. Think of large-scale hydro as the last generation of renewable power—big dams, siltification, habitat destruction, and anadromous fish interruption (think of the lost salmon and shad runs). Now compare that to what human beings can do today.

why we believe the best solution is a humble approach combining small solutions that add up to something huge.

Wind power was a renewable that seemed flawed early on but is now fixing its problems step by step. Wind power had the advantage of not massively reconfiguring our terrain at the start and *then* discovering the downside. That relatively smaller profile allowed wind power to adapt, fix itself, and grow.

Wind power, of course, is nothing new. Windmills have been used for centuries to grind grain on Mykonos, drain the polders of Holland, or pump water from wells into fields in North America. Before rural electrification in the United States, tens of thousands of small electric wind generators dotted the rural landscape. In the 1970s and '80s, the U.S. government funded the development of large-scale wind turbines, as did other countries such as Denmark, Germany, Spain, India, and, later, China. Now there are many manufacturers of large-scale wind turbines in the market, including big companies like General Electric.

The market for wind blows hot and cold based on local energy pricing, available tax incentives, market production, transportation costs, and so on, but generally, it has been a fast-growing renewable energy sector. Since the 1980s, the cost per kilowatt-hour of wind has dropped 80 percent; it is approximately 2¢ cheaper per kWh than coal-powered electricity on the U.S. market as of June 2012. And that's without accounting for the cost it saves our system due to decreased carbon dioxide emissions. About 24 percent of Denmark's energy needs reportedly are met by wind, and the country has plans to grow that to 50 percent by 2020. Serious wind development is happening. It has all the hallmarks of a new industry with ups and downs, but it is clearly here to stay—even with cheap natural gas coming from hydraulic fracturing (fracking) in shale formations.

But for many people, wind power doesn't seem like a clean renewable because it can create visual blight on formerly beautiful country landscapes. Where wind turbines crowd, say, the Tehachapi Pass in California or the Nantucket Sound in Cape Cod, some observers think they might as well be looking at big, unattractive power plants.

Many individuals and government bodies have begun to address this eyesore problem.

The University of Maine is developing the engineering for large floating wind turbines 20 miles offshore (out of sight) that would produce the energy equivalent of 150 nuclear power plants. Maine could have a new cash crop to add to its famous lobsters—clean energy. Minnesota recently passed legislation to provide incentives to individual wind turbine owners, residents of the state, who do not own more than two wind turbines. In other words, you can win tax breaks on property—the purchase of the wind turbine itself—and a business tax credit if you generate wind on a small scale. This sort of legislation could help every Midwestern farmer buy a source of renewable power for his or her work, generate income by selling up to the grid, and fashion a distributed generation network appropriately scaled to the other human activities in the locale.

Dan Juhl, who understands farm life from his own childhood in Woodstock, Minnesota, has been creating profitable business arrangements to encourage this kind of small-scale wind-plant cooperative. Often big utilities and energy buyers enter a community to negotiate the cheapest price for land and end up pitting one farmer against another. One farmer becomes the winner and all the others get left out. Instead, Juhl, through his Juhl Wind Inc., brings together a group of farmers to invest in a common future.

For a set rate of return, an outside investor injects significant capital to fund the larger construction costs. As the tax incentives and revenue accrue from producing and selling electricity, the investor is paid out his or her return. In ten years, when the investors have achieved their goals for financial return, the ownership of the wind turbines transfers to the farmers, who can use and sell the energy produced. The wind turbines become the capital real estate of the local residents, producing "currency" for the perpetuation of jobs and benefits to the local community for decades.

The result? Wind turbines dot the Great Plains, local family farmers earn enough for their mortgages and their kids' college educations—and a new industry, renewable power, is created *in places where we need power.* The investor is happy too, because the guaranteed return has been paid out in full. The dispersed wind turbines make a more pleasant visual for the neighbors. Wind is the new cash crop.

Another issue with wind has been how to store the energy produced, not just sell it to the grid. This issue has sparked creativity and innovation. One of our favorite small-scale proposals involves using wind turbines to help solve a pressing rural problem. In farming regions, school authorities have been having difficulty affording school buses to pick up children each day. The farms are too far apart, and the fuel gets expensive. But of course everyone wants their kids to be able to live on the farms, for the parents not to have to trade in their lives as farmers simply to be sure their kids can be educated.

Juhl's solution is to install small wind turbines, an optimized design, at community centers in the Midwest and use these centers to power up electric buses. These vehicles require less maintenance because there is no need to change spark plugs or oil, as with internal combustion engines. The buses deploy in the morning and the late afternoon. The rest of the time they are sitting idle. But what if that time were optimized? What if, when parked, waiting, the buses were getting charged by the sun and by the wind?

The wind would actually be healing the community. The children could still be at home on the farms, and the buses could fetch them for their school day. The community would have optimized around a local resource for concentrated energy instead of farmers sending their money far away to the Middle East looking for fossil fuels that are insecure and might inspire even military intervention. Or instead of spending their money on nonlocal sources in Pennsylvania or Alberta, they could collect the clean, free, abundant energy flying right overhead, ready to pay for mortgages and college tuitions. No one has to leave the farm to protect a way of life.

Letting Industry Strike the First Blow

Another problem wind originally had was finding the investment capital to get the turbines built. But many companies buy CO_2 offset credits for their power uses, either out of their own concern for the environment or as an investment or PR strategy; investment in wind has now become one option for how to do that.

In 2008, Steelcase, the office-furniture maker that produces the Think chair, invested in a Texas Panhandle wind farm to help offset carbon releases. The renewable generator that was created—Wege Wind Farm—powers almost 3,000 homes and businesses with clean energy, and Steelcase receives 20 percent of the farm's carbon offsets for its own production needs. The company gets the satisfaction of knowing it is making a specific investment in wind power for a community that needed such an energy source. It's a CO_2 offset, but the real key is that it's creating what we call additionality—i.e., creating more of a good thing, in this case creating a new renewable power system, instead of merely supplementing the cost of existing renewables.

Other companies are taking these steps too, following aspiration and commitment with action. Herman Miller, another major furniture manufacturer, has cost-effectively implemented wind power solutions that support the manufacture of all their Cradle to Cradle products. These kinds of initiatives are the beginning.

Another Way to Offset CO_2 ... and Make Energy

Here's another energy solution: A cow produces enough manure a day to provide more than 2.4 kilowatt-hours of energy—in other words, the power needed to light one incandescent bulb to burn all day, or three CFLs (compact fluorescent lights), or five LEDs. Biogas is a near perfect example of upcycling. It is the methane and carbon dioxide produced by biological matter as it decomposes, which then is burned for fuel. If the cow manure and decomposing vegetation or other organic matter were contained in a biogas plant, not only could the off-gas turn into power, the methane and carbon dioxide that would otherwise be released into the atmosphere would be contained. Methane is estimated to be more than 20 times as potent a greenhouse gas as CO_2, so that containment is a great asset.

The sources for biogas material are everywhere on our planet: from gardens to cemeteries, from wool to pet droppings. We are seeing landfills that have been capped and the methane is being collected and used to provide power—electricity, for example. Companies such as Axpo Kompogas in Germany, a CO_2-neutral fuel provider with more than 50 facilities around the world, are showing the way with their fermentation processes. Between 600 and 1,000 kilowatt-hours can be generated from one ton of waste (approximately enough to power a typical energy-inefficient house for 38 days), and nearly a ton of natural fertilizer is created as a by-product. The United States alone produces 220 million tons of garbage a year. Imagine the power that is possible.

Clearly, biogas could not sustain all our energy needs (as our needs are currently designed), but it would certainly assist. It's readily available and right in front of us, it can be optimized around greenhouse gas concerns, and it adds no foreign processes to the natural progression of material into soil. As you will see in the next chapter, human beings can continuously benefit from soil enhancement.[8]

Dam: Upcycle in the Desert

Okay. We have discussed various renewables and their attributes. We have discussed the grid. We now want to invite you to engage with us in an evocative, big-thinking exercise about how to take all we have discussed in this chapter and power something huge. How about the United States?

Let's take a fresh look at the Hoover Dam in Nevada, America's most famous hydroelectric dam, responsible for a major WPA job creation program and for enabling Las Vegas to exist. Remember jobs and power again when we get to the end of this chapter.

If one had the need for a certain amount of power—and let's presume that humans could optimize the demand side (is it possible to imagine an energy-efficient Las Vegas?)—then how would one, in the desert, deliver that energy? Would you block up the Colorado River, which is how we got Lake Mead, the body of water that was created when the Hoover Dam was built? Because of extraction for cities' water needs, as well as evaporation, the Colorado River rarely reaches the Sea of Cortez anymore. The delta claims only 5 percent of the wetlands it had previously. How many people lived in the delta of

---

8. We could be looking at methane extraction from our garbage, not just capping landfills, as is often done. In an optimized system, we could extract methane immediately from our garbage for practical purposes—heat and electricity—and then use the resulting materials for soil amendments. Instead of leaking methane over centuries in bits and pieces, we could harvest it. As it is now, well-meaning people making compostable packaging might be seen as adding to our methane-release burden—another ill effect of "good behavior."

the Colorado in Mexico before the Hoover Dam was created? Was that dam designed with these consequences in mind?

The people who love the Hoover Dam probably prefer not to imagine a world without it. They say it's a massive economic development area because of powerboats and water skiing. But one can build recreational lakes without having to dam the Colorado.

Instead, what if we switch our thinking? As an imaginary exercise, and just to get a feel for how amazingly powerful the sun could be, consider exactly how much desert area covered in solar collectors would be required to power the entire United States: 140 square miles. How big is that? (We will discuss distribution later on.)

Get in your car. Start driving.

At some point in the desert you will start seeing a gleaming field of solar collectors on your left side. They spread on, as far as you can make out in the haze of the heat mirages. They are a fairly daunting spectacle. You keep driving at the speed limit. About two hours later, you come to a corner of the field of collectors. You turn left, following their edge, driving south into the desert.

The gleaming field stretches deep to your left here too. About two hours later, you turn again to the left to follow the southern edge. Two hours later, one more left and you get back to the point where you first saw the collectors. You have just driven around a field of solar energy collectors that can power the entire (mostly inefficient at this point) United States of America, a whole country.

That square is the equivalent in area, relative to the total area of the United States, of a beach towel on a basketball court.

Now get out your scissors.

Cut up the beach towel into tiny pieces. Spread it around. Connect them with a thread.

How about putting some of those pieces outside Las Vegas, or even on the rooftops of Las Vegas? Making them solar? In a new Las

Vegas development program, if developers wanted to build houses, office buildings, or casinos, they could install parking lot and other solar collectors to meet the new demand.

What if we used existing highways for renewable energy distribution? Imagine ribbons of road running through the desert covered with lightweight shading devices made of solar panels, or with the panels stationed adjacent to the highways in the public rights-of-way. Instead of drilling for oil and gas on public lands and in national preserves, where few benefit and we put more carbon into the atmosphere, we could mine solar or wind on public rights-of-way for energy and jobs. What if we optimized the existing industrial landscapes? You could obviate the need for massive doses of fossil fuel or nuclear power plants, and the system, because it would be dispersed, would distribute power where it is needed. If you crunch the numbers, you'll see that this really would solve our energy needs.

In Europe, Michael has encouraged putting wind turbines within large power transmission towers, since these towers already represent public infrastructure visible in the landscape; the urban blight issue would be contained, since these areas are already in use for industrial purposes. Such proposals would also address grid issues, since the tower locations could be fully optimized for electrical transmission and power generation. Regardless of whether or not these technologies are ready for installation today or will even make practical sense, this is the kind of thinking that leads to innovation. For innovators, it's a worthy stimulus for dialogue.

Bill has been developing a similar idea whose time may have come: using public lands to help America achieve energy independence. Let's call it upcycling Amtrak.

We begin by considering infrastructure we already have. If we look at the Amtrak train system, we learn it was subsidized at the level of $1.4 billion in 2011 alone. What if on the land right around the train tracks, which already has been secured as public rights-of-way, renewable power systems were installed? Our team ran those numbers and it appears that the area available for potential use in generating renewable energy would be so significant that Amtrak could be *the* major contributor to the U.S. power requirements in a distributed fashion, even using only solar collectors.

Just imagine railroad lines running across Kansas with solar collectors on the rights-of-way on both sides of the track. What a terrific resource and what a great location, since it's ideally positioned for maintenance and transmission using the infrastructure that is already there.

Suppose a map of such a grid were shown to Amtrak officials and they were told, "You have 14,000 miles of easements that could be used for solar collectors." Amtrak, almost overnight, could provide the real estate for the distributed large and small power utilities all over the United States, along with thousands of local jobs all across the country. Why couldn't Amtrak be financed, not subsidized, by multi-megawatts of power instead of billions of tax dollars? Not to mention the opportunity to run solar-powered electric trains.

Since this grid would be able to uptake energy, all manner of localized energy production could be integrated and employed. Add wind turbines where appropriate. Add the methane from our garbage. Add biogas, gastrification of organic materials from our farms. We are not talking here about ethanol production for fuel; we are talking here about biofuels. We are talking about normal life becoming productive without the strange need for ongoing subsidies.

People just need leadership, guidance, and optimization from an engineering perspective to get such a project done. Society would get jobs, mobility, clean energy, and distribution. We would get to send a signal everywhere we go that the world is getting better. Every mile you travel, every train ticket you buy, becomes a story of regeneration, of the beneficial opportunity to mine public land for public benefit.

This scenario can evolve a step further. How about upcycling our treatment of the United States border with Mexico? Imagine that we would like to send a sign to our friends south of the border in Mexico. What kind of signal do we send now? Fences saying, "Stay out." Instead, what if we created a giant welcome mat to the future? Instead of spending hundreds of millions of dollars digging holes to put up only marginally effective barriers, what if we dug holes and built solar collectors, a giant sculptural ribbon—like the artists Christo and Jeanne-Claude's *Running Fence*? The United States could create an ad running from Imperial Beach, California, to Brownsville, Texas, announcing that it is a renewably powered country, to benefit the world, and is happy to share innovation.

We could help Mexico build its solar industry too. We could have thriving border communities where we meet and share. Upcycling relationships. Renewable friendship.

Ask Not How Much the Grid Can Give to You but How Much You Can Give to the Grid

Now that we've talked about energy sources, we want to discuss the system of delivering that energy. Some people talk about getting off the grid. As you can tell, we are quite interested in *contributing* to the grid, from a diverse and distributed perspective. We are not suggesting everyone run away and be independent. We want to celebrate interdependence.

The grid can be optimized to take advantage of the possibility of local power production fed into the common system. The concept of distributed energy is nothing new. The Jacobs farmstead wind turbines that dotted the Great Plains of the United States or the windmills of Europe were locally distributed energy systems. They were small and they worked, or the farmers would not have bought them.

The difference is that when people added transmission, the system moved toward centralized power because of the efficiency of concentrated sources and AC distribution. In conventional electric power production, energy is created at a base-load station, with peaking power plants backing up production for high-need fluctuations. When power is transmitted along the grid to your house or business, it whispers away due to line resistance. The farther the electricity must travel, the more energy is lost, up to 10 percent.

The power goes to a step-up transformer, which, if it is working properly, doesn't lose more than another 2 percent of the energy.

Then it runs along transmission lines to a step-down transformer. The purpose of a substation is to step down the power load so it doesn't explode your toaster oven. Again, a loss of about 2 percent. It moves on to a subtransmission customer, to a primary customer, and then on to you.

We could say the total line loss is about 14 percent. For every 10 tons of coal burned to make your electricity, 1.4 tons are lost just to move it around over long distances. It's suboptimal. And you, the customer, have to pay for it.

But if you have a solar panel on your roof, there's no distance. There's no security threat, because the whole system is dispersed. The power you collect can be used or even, in 43 states, sold into the grid when you receive an excess of productive sunlight. As society moves into the future, the ancient practice of a distributed unconnected system can be integrated with the connected system and then move into decentralized generation within a connected system—upcycling the grid. The whole population could be not just consumers but energy creators.

Hundreds of policy makers and entrepreneurs are working on this decentralized model, and it appears to have immense potential from all sorts of economic, sociological, practical, and security perspectives. Walmart is committed to becoming 100 percent renewably powered as soon as it is cost-effective and even looking at ways of becoming a power wholesaler in certain markets.

Clearly, this will cause disruption—you could even call it Schumpeterian disruption, creative destruction. Existing utilities that are providing base-load power can feel disrupted by renewables coming on in new business models. A new system of generating and transmitting power will require society to develop the frameworks for this to be optimized, which may be difficult with a lot of incumbents fighting that change, especially since the incumbents are very entrenched, very organized—might we say, powerful (pun intended). (Imagine Thomas Edison's undaunted courage and optimism taking on the whale oil or gas lamp lobby.)

It may sound herculean to transform existing electrical generation and distribution systems, but the creation of the federal highway system under Eisenhower is an astonishing undertaking if you stop and try to imagine it all at once. The Internet is astonishing if you try to consider it as a whole. How could it be invented in its scale to date? And yet it has rapidly come into being in recent times. How quickly can these things happen?

The Little Lightbulb That Could

Remember, in our energy formulation, how we said we want to consider the whole circuit, from source to grid to where and how the energy is used? As we described before, when Edison invented the lightbulb, he also understood the need for electrical energy generation and the grid. We have talked about the generation of energy and the grid. But if we don't consider how we enjoy its benefits at home, or in the office, or at the factory, the entire energy discussion is left unfinished. If we optimize the design of our buildings, appliances, and consumption patterns, the energy saved is potentially so huge it could make us feel as if we have just (to borrow yesterday's enthusiastic phrase) struck oil.

We could talk about a refrigerator or a house or a train or a city, in rethinking the appliances of modern life. But let's return to the lightbulb. It's the perfect device to consider when aiming for optimization since it is so small, so important, and yet could make such a huge energy difference.

Traditional incandescent bulbs, the very lion of a new idea, have gotten old. They are now a problem for modern society because they burn off too much energy in heat and not enough creating light. That's why countries from the Philippines to Argentina to France, Finland, and many more are banning the production of inefficient bulbs.

But as we discussed earlier, compact fluorescents contain mercury. CFLs reduce a negative effect in one area of the system (energy demand) but cause a problem in another (materials toxicity). So while humans are being "less bad" in terms of power usage, they end up with many, many bulbs in landfills releasing mercury—a toxic rare-earth metal that does not safely circulate in biological systems. (It could, if safely utilized and managed, be *prized* as a rare and valuable "technical species," as a component of safely recycled technical products.)

In response to new laws to create more energy-efficient bulbs, many companies are innovating on Edison's design and "form factor"—the A bulb—to provide more effective and efficient light per watt without mercury. LED high-performance bulbs can retain the basic shape and form of a standard bulb while configured to cast an extraordinarily flattering and enjoyable glow. LEDs also have the advantage of generating bright illumination. While LEDs are more expensive at the moment than conventional bulbs or compact fluorescents, it is expected that the price will come down to compete with CFLs—they dropped from $36 to $18 per thousand lumens in 2010 and are expected to decrease to $2 in the next few years—and we think they will then be popular. A bulb will last for 10 to 15 years, and every part of it can be designed to be safely reusable in technical cycles. We are working with lightbulb companies as part of their product creation programs to strategize so components can later become anything from a bicycle part to a nutrient in the biosphere, or even . . . another lightbulb. Imagine if we took this a step further, and the bulb materials were products of service that you leased from the maker—you would have light overhead rather than bulbs in the trash. What an idea!

With LEDs in hand, we would now like to look at a whole upcycle theoretical model, from power source to power need.

In a fanciful analogy, let's return to the Hoover Dam: If the surface area of Lake Mead behind the dam, which produces 4.2 billion kilowatt-hours per year, had solar collectors floating over it, the collectors would actually produce 10 times the energy the dam produces. (As a bonus, the floating solar collectors would reduce water evaporation.) Now, while the boaters may wish the whole lake were still set aside for boating, we can renegotiate the shared water space later. Right now, we wish to float collectors in this imaginary model.

In 2010, the estimated energy needed to power residential and commercial lighting in the United States was 499 billion kilowatt-hours. How could we reconceive that need with far fewer kilowatt-hours and from renewable power? (Once again, companies such as Google, Walmart, and Procter & Gamble, looking to save money for customers and help them live better, are doing these things. There are a great many examples of leadership . . . and plenty of opportunities to follow.)

The numbers we are about to give you are rough—we are not accounting for the different savings of different wattages or the possibility that in 2010 a substantial portion of the lighting in use might have been compact fluorescents—but here's the conceptual model. Let's say this lighting demand across the United States came from 100-watt bulbs, and you switched out all the lighting for the new positive material LED bulbs, whose energy demands are typically 20 percent by comparison. The dramatic reduction in wattage per light fixture in America could result in an astonishing energy drama. The country's energy needs for lighting would diminish to 99.8 billion kilowatt-hours.

In terms of Hoover Dams, and just for effect, if we add in the supply side of this discussion, with the tenfold increase in energy production of a *solar* Hoover Dam/Lake Mead, you would need only 2.4 solar Hoover Dams to light up the whole country. In our current system—with suboptimal lighting systems—you need 119 Hoover Dams.

With redesign, the whole required energy system shrinks. What we thought we needed for energy might be a whole lot less. Isn't this amazing?

Is this idea overly dramatic?

Well, let's think a minute. The Hoover Dam was considered an engineering miracle of its time. Was that overdramatized? Wasn't that a dramatic moment, a project as ambitious as that? Was the moon shot dramatic? The federal highway system? Would you sit back today and say, "You can't do that," because building a highway system all over the United States, six lanes of traffic in every direction, is too big to consider?

Is it too big? Or are human invention and daring too small?[9]

Our theoretical model brings up opportunities for all sorts of lightbulb jokes:
Q: How many Americans does it take to change a lightbulb? A: 311 million.
Q: How many libertarians does it take to change a lightbulb? A: None. They wait for the market to do it for them.

Take this another step: What if society upcycles the energy question "How can the system create adequate energy?" to "In a world of abundance, what does society do with its excess energy?" What's next? How might people upcycle systems so that not only does wind equal food, but we could also fruitfully store energy *in food*?

Soil
Not Oil

Chapter 4

Food as a battery—that is what we would like you now to consider. But before we get to the full expression of that proposal, we need to review exactly how batteries function, so you can appreciate the beauty, and potential innovation, made possible by thinking through this metaphor.

Batteries are not storage containers for electricity, as one might assume. They don't provide power because somehow someone pumped in the electricity and locked it in, and now it's ready for use. Instead, they contain the *potential* for an electromagnetic reaction, which, if engaged, creates power. The battery consists of a negative solution (the anode) and a positive solution (the cathode) separated by the ions of the electrolyte. The extra electrons in the anode want to move to the cathode, but there is no path through the electrolyte between them.

When a wire connects the negative end to the positive end of the battery, the electrons can flow through the wire, seeking their harbor in the cathode. These free-flowing electrons, in the middle of that path, power your flashlight or start your car.

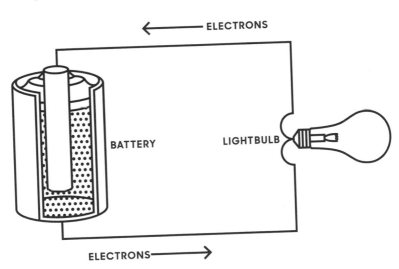

The beauty of a battery is that it is *potential* energy, ready for your use, when and where you need it. Should the battery run out of charge, its power is recharged by reversing the process, forcing the electrons from the cathode into the anode. Then you can start again using your battery to provide electricity.

Now think of how humans conventionally create energy. We burn fossil fuels—i.e., carbon-based organic compounds (as we have said earlier, fossil fuels are ancient organic compounds)—and inadvertently turn them into carbon dioxide, among other things.

Photosynthesis is an electromagnetic reaction that frees electrons from water to turn carbon dioxide into organic compounds.[10] It is the reversal of the burning of fossil fuels. It is recharging the battery. It is recharging our power source. If people don't allow the recharge of that battery, the world can't recapitalize.

If one looks today at our organic battery, this biosphere, which has provided all the energy that people have used for their needs for millennia (the fossil fuels in coal and oil; the biofuels in wood), one might begin to understand the importance of recharging. Human beings have every reason to want to do so.

Get Down to Earth

Let's look at the common worm. As a worm makes its sinuous way through the soil, it aerates, tills, plows, and fertilizes. Of course, it doesn't intend to do these things, but it seems to have been designed,

---

10. The photosynthetic act claims six times the amount of energy in terawatts that humans currently use. Plant life is greedy with its energy needs, greedier than humans are, and yet because of the nature of the energy source for vegetation—the sun—there is no reason to complain. Not only that, the photosynthetic process usefully absorbs energy and stores it, often usefully for our potential need, either as food or as fuel.

by nature, to have beneficial effects in the course of every single thing it does.

Worms are avid consumers. They eat their own weight in food each day. Yet they are enormously helpful to ecosystems (our use of "yet" indicates how much people have come to associate "consuming" with destruction and waste, which is certainly not the case in nature). Worm castings—what they leave behind—are "waste" only for a moment before they become "food": These castings are rich in nutrients, extremely rich—they contain higher levels of nitrogen, phosphates, and potash than the soil around them. The lowly earthworm is one of the planet's most valuable creatures (and apparently one of Darwin's favorite organisms).

Compare this highly effective and evolved interaction with soil to humanity's most recent interactions with soil. Humans have the capacity to be similarly effective as earthworms. One way is to add nutrients, and we could easily do so, but so far, for the most part, we aren't.

How can we do this? The history of the development of the technical battery over time has been one of experimentation with various substances to sustain the longest charge and promote the most powerful chemical reaction needed to create the flow of electrons; to reduce the size and cost of the battery while optimizing the duration of its energy output; and to create specific batteries for specific products and needs.

To translate this to the earth battery: We might create farming techniques that sustain the longest period of productivity, augment the soil for optimal plant growth, utilize soil in the most compact way, and diversify the design of that growth for different locations.

Right now in human history, we have designed and implemented a system that puts us in danger of expending our earth battery. Carbon is not treated as a valued asset by human industry. People do not feed the soil. Since the founding of the United States, the country has by some accounts depleted 75 percent of its topsoil. Most of this loss is caused by now questionable modern agricultural techniques—monoculture (growing one kind of crop year after year, so the same nutrients are siphoned out), overtilling (which encourages topsoil to become airborne and erode), and salinization of soil caused by overwatering and overuse.

One hundred and fifty years ago, the Iowa prairie had 12 to 16 inches of topsoil, as well as the carbon stored in the deep roots of prairie plants, which were as much as 15 feet deep. Now the topsoil is down to 6 to 8 inches. Soil production takes significant time; it can require from 100 to 500 years to create one inch of topsoil. With those kinds of numbers, human beings have little to no hope of catching up.

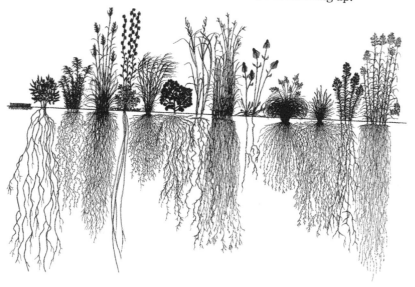

We are frittering away our future food. Some estimates for the United States show that 6 percent of wheat and corn production is lost for every inch of topsoil that disperses into the air or water. Or to put it in other terms, the United States is said to lose $125 billion worth of topsoil a year.

The problem is occurring around the world. The United States loses topsoil 10 percent faster than it can replenish; China and India are at rates of 30 to 40 times faster.

The quantity of loss isn't the only problem. People are also depleting the richness of the soil that remains.

Norman Borlaug, the agronomist known as "the father of the green revolution" who won the Nobel Peace Prize in 1970, came up with revolutionary ideas about hybridization to optimize grains for higher yields. The Nobel selectors credited him with saving more than a billion people from starvation. But those green revolution concepts have now inspired industrial farming to escalate hybridization and genetic modification to the point that they are selling an herbicide to kill weeds and then crop seed that can resist the herbicide. The farmer is buying at least two different products—seed and herbicide—from the same corporation. Farmers have also become more dependent on soil additives, such as phosphate, which are customarily mined, requiring the farmers to go far afield—and certainly far from the field, even to distant lands—to maintain high local yields.

The green revolution has been enormously productive, but its focus has essentially been on gleaning energy from the battery without considering the optimized design of the organic battery, for how to maintain the charge. We think that human beings can be doing more to recharge their local earth.

Soiling the Planet: Give Back

The second thing we can do for our earth battery is optimize the soil to encourage electron exchanges. We can improve plant life's access to needed nutrients in soil.

We are extremely interested in the alternate green revolution launched when Sir Albert Howard published his seminal *An*

Agricultural Testament in 1940. He was an agronomist sent by Britain to India to inform the farmers about Western farming techniques. To his surprise, he found the Indian farmers functioning quite well. Their agricultural systems were focused not just on optimizing specific plants but on maintaining soil health—and, more specifically, on devising systems to sustain the microbacterial matter in the soil. One example: The Indian farmers were able to return difficult-to-break-down straw to their soil by putting it on their roads, crushing it with farm wheels, and mixing it with manure. Howard's insights introduced the idea of modern composting and led to the beginning of the organic agriculture movement.

These advances play out today. As just one small example, Gary Zimmer, of Midwestern Bio-Ag, advises more than 3,500 farmers working approximately two million acres of farm—primarily in the Midwest, but reaching as far as Idaho and Pennsylvania—to build productivity from the soil up. He looks first at what the soil on each farm needs in biological nutrients. His techniques may create yields slightly lower than those created by the big agricultural companies, but the cost in soil amendments is also lower, so the profit can be higher. For farmers around the world, the idea of lower costs with higher profits probably sounds appealing. It's common sense.

Our point is this: Many people believe that the next green revolution will be an offshoot of the Borlaug revolution—that it will come from optimized and modified seeds and plants, and certainly those developments will continue. But we believe the next green revolution may come from the soil. In other words, it may come from people trying to execute the optimization of the battery—the way the earthworm does. And all of this will be further amplified by greenhouse techniques such as hydroponics.

Phosphate: The Next Fossil Fuel War

Phosphate is one of the key ingredients in soil, in how the earth re-charges itself. Plants require phosphate to grow. Animals, including ourselves, need phosphate for bones, teeth, and membranes—and we get that mineral from our food. Plants, clearly, get their phosphate from the soil, and in nature's system they would return the phosphates to the soil when they die and decompose (or are redeposited as animal waste).

But humans have been implementing less-than-optimized practices. Through farming, people are removing high levels of phosphate from the soil—the plants take up the phosphate and are then carted off, leaving no remnants behind to "reseed" the organic phosphate.

Human beings also put the phosphate available in soil out of reach by overwatering. This does not dilute phosphate; it causes it to bind to other elements before the plants can uptake the phosphate in a form useful to them. In fact, the phosphate binds with so many other elements—silicates, carbonates, sulfates, and the like—that it can be a tricky nutrient to add to soil. One gardener has compared it to throwing a monkey through a jungle. The monkey's tail and arms and legs catch on so many vines and tree limbs that it can't get through.

The current solution to reintroduce phosphate to the soil involves dumping mined phosphate onto fields. But because phosphate links to so many other elements, it easily washes out of the soil and into groundwater, where it leads to the high nutrient content in lakes and rivers that subsequently creates algae blooms, killing off fish and aquatic plant life. Mined phosphate also tends to include radioactive elements, such as uranium, radium, radioactive lead, radon, thorium, polonium, and cadmium, because these are inescapable trace elements in phosphate ore extraction.

Plus, there is a geopolitical conundrum in the use of mined phosphate. The top two exporters of phosphate in the world are the United States and China, followed by Morocco. But in 2010, China, recognizing the importance of phosphate to its own agricultural needs, slapped a temporary 110 percent tariff on exporting phosphate at the cusp of the spring planting season. That left the United States exporting its dwindling supply. At current rates, the United States' supply is estimated to be depleted in 30 years. That means the United States will be dependent on imports from Morocco or China—which could get expensive as tariffs fluctuate—much as nations are dependent on imported oil.

To come up with a solution for this phosphate requirement might sound like a daunting challenge, but the solution is not out of our reach. In fact, we all get to the bottom of it every day.

Before we talk about the solution, though, we need to introduce a concept that will prepare you for the revelation: Sometimes in order to make a new idea acceptable we also need to upcycle the language.

Take, for example, an impressive effort that San Diego, California, began in 2007 to study "recycling" sewage water to address the city's very real water shortages. In the absence of an official name for the operation, journalists started calling the reclamation effort Toilet to Tap.

Now, it's hard not to see the yuck factor in that phrase. And, unsurprisingly, San Diego citizens balked at the idea of consuming their own wastewater. It didn't matter that the recycled water, after being sent through the purification process, was in fact cleaner than the water San Diego residents were currently drinking. No one wanted to think about toilet water in his or her drinking glass. Sydney went through a similar experience during the droughts there.

Singapore, on the other hand, had its water-recycling pitch well-tuned right from the start. When conducting feasibility studies on technology for reclaiming water, they called the project NEWater. With that nice, refreshing term, citizens were inspired to take pride in the idea that they were being endlessly resourceful; NEWater now accounts for 30 percent of the country's water needs. Upcycling allowed Singapore to stop importing water (from Malaysia, which they had been doing for years, despite constant political friction), bringing greater safety and security.

Keep that issue of language in mind while we reveal something that we think would go a long way toward assisting how humans interact with nature.

In the Western world, for more than a century, people have been misled into thinking that our "waste," what we flush down the toilet, is somehow toxic, that it cannot be worked back into the natural system, that it cannot be used as compost for growing plants. This is not true. Your waste is manure as helpful as any other manure on the planet; it just has to be handled correctly. Your urine, over a 24-hour period, contains half the phosphate you will need to consume in a day for healthy bones and teeth and tissue.

For millennia, people understood how helpful our own "emissions" could be. When Bill was a young child in Tokyo, he would hear the farmers coming through the streets when everyone else was in bed, using their "honey wagons" drawn by buffalo to collect the night soil (human waste gathered from cesspools and privies, for use as manure). At that time, people could buy such "waste."

The Japanese required intense cropping, and where else could they get their phosphate? It doesn't just rain phosphate.

The Japanese were sensitive to the handling of pathogens, and they knew how to compost the night soil before they used it on plants. But the way humans treat "waste" now is to call it sewage and chlorinate it, then dechlorinate it with sulfur. Some systems use ultraviolet disinfection. All these processes require tremendous energy loads, about 4 percent of the United States' total electricity expenditure. And the "waste" still ends up going back to pollute the larger water system, along with runoff from septic systems.

We could change what is essentially grave mismanagement. Humans can upcycle sewage. Stop thinking *sewage* and start thinking *nutrient management*. Stop thinking *ugly, smelly liability* and start believing the old adage *money doesn't stink*. In fact, that expression from the Latin, *Pecunia non olet*, came from the Roman emperor Vespasian, defending the unsavory nature of his tax on public urine. Roman citizens bought urine to tan their leather and clean clothes and were taxed accordingly. When Vespasian's son expressed his repulsion, the emperor held up a coin and asked if it smelled bad. The son replied that it did not, and Vespasian pointed out that it was earned from urine. Money doesn't stink.

As with Singapore water, the yuck factor will dissipate by not only changing a term but also realizing the immense profitability of reusing our biological nutrients. We are encouraging cities like San Francisco and countries like the Netherlands and Sweden to convert sewage into valuable products like phosphate and nitrogen.

A company in Vancouver, British Columbia, is developing ways to recover phosphate from human waste. An engineer at the sewage company had been studying the problem of waste pipes clogging due to the crystallization of minerals in the pipe—a liability. The engineer attempted to get the crystallized mineral out, but this proved very hard, literally, because the minerals were stonelike. So the engineer came up with a mechanical device and a small chemical intercession. The mechanical device created a vortex, a swirl, spinning the water so the minerals wouldn't cling to the pipes. What happened then? The minerals came out as pearls—of phosphate.

These pellets of magnesium ammonium phosphate are known as struvite, and for farmers they're ideal because they release their nutrients slowly, taking about eight to nine months to fully dissolve. They feed into soil at a pace that plants can digest. And the farmers don't have to keep laboring to add phosphate since, for eight to nine months, they know the fields have their fill.

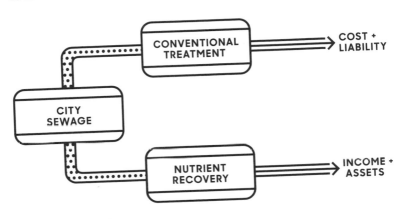

The sewage treatment plant, which had defined sewage as a problem to be contained, could now become a nutrient management system, capturing phosphate to feed soil. Yesterday's cost became today's coin.

Let's look at what this kind of transformational thinking can mean: Many cities are on large bodies of water and therefore need to think about how sewage treatment plants could transform into nutrient management systems. With struvite harvested in this new way, the capital cost of deploying the system can be repaid in three to five years, and then the city begins making money by selling the phosphate. This approach to nutrient capture also applies to nitrogen.

A city could also supplement that business by making biogas from the methane emanating from the sewage, thereby reducing the release of greenhouse gas while producing energy for sale or to power other operations. The cost to the city of providing a sewer plant has been upcycled into a profit generator.

We have been presenting this idea to locales for years, and we have been delighted by the uptake. Instead of farmers buying slightly radioactive phosphate from Florida, with point-source pollution (a confined and identifiable source, in this case the sewer pipes directly dumping pollutants into a bay), the farmers are receiving a high-quality fertilizer from nearby, and it's slow-release. Some of the phosphate crystals can take a year to dissolve—eliminating non-point-source pollution (an uncontained source, such as storm-water runoff or fertilizer washout).

All of a sudden the economic vitality of the nutrient-polluted Chesapeake Bay, for example, can return with its seafood industry, with its jobs and culture. It can come clean—come back—for good. The city can start making money. Less money spent on transportation, less money spent on environmental cleanup, more money earned by selling struvite and nitrogen and methane.

Toronto is trying a different method of upcycling goods formerly discarded because of the yuck factor. Residents are encouraged to throw disposable diapers full of poop into green recycling bins dedicated to accepting all compostables. The program also takes other materials usually designated undesirable, such as kitty litter, soiled paper, and sanitary products.

At the composting center, the machines sift the materials, then blend and anaerobically digest them while the biogas is captured for energy use. Society can continue to debate the environmental cost of using disposable diapers, but if people are using these diapers already, why not use them constructively, effectively? The diaper can actually help rebuild topsoil. Within seven months, the composting process is complete and the city gives the soil back to park managers and residents for gardens. Michael has shown that, properly deployed, each baby's diaper use could provide the moisture retention and nutrients for starting more than a hundred trees in a desert.

We are also seeing the use of flies for composting and animal production. The maggots decompose and convert organic vegetable matter or flesh from slaughterhouse scraps into the amino acids and proteins needed by fish and chickens for their diet. In this case, something considered garbage becomes an animal resource, which then can be converted into protein desired by a significant portion of the human population. There might be an attendant yuck factor involved with breeding flies for the food chain, but remember, free-range chickens love insects, and as for fish . . . ask any fly fisherman.

And it is surely less yucky than feeding chickens arsenic to plump them, as many poultry farms currently do.

How to Make Anything a Battery . . . an Ever-Resourceful Battery

First, let's look at battery-design optimization in terms of how to hold charges longer. For this earth battery, the emphasis needs to be on how to design agriculture to increase sustainability and growth with optimized interventions. That's where permaculture, the development of agricultural ecosystems intended to be long term, comes into play.

One of our favorite permaculture experiments, which started in 2002, involved a 10-acre stretch near the Dead Sea in Jordan. This particularly arid stretch of land, with its salted soil baked up to 122° Fahrenheit in August, had frustrated farmers for millennia. In modern times, either they grew their crops under plastic or they drowned the area with pumped-in (and precious) water to push the salt 20 feet down (this ultimately damages the soil for 1,000 years). To encourage what little growth they could get, farmers poured fertilizers and pesticides onto crops. (In *Cradle to Cradle*, we talked about how cradle-to-grave design often relies on the use of brute force to get the job done. Brute force is not necessarily an effective or efficient or elegant tool to accomplish a task.)

But Geoff Lawton of the Permaculture Research Institute in New South Wales, Australia, thought the farmers might get better results if they worked more holistically with the elements provided to them. They might create a real oasis.

Lawton's first step was to use the rainwater to its fullest. He and his team dug curving swales—ditches—to collect the small amount of rain that falls in the region in the winter. The ditches managed to catch about 250,000 gallons of water in the winter and slowly leach the moisture back into the soil. Then Lawton created mounds on either side of these ditches and piled waste from a nearby organic farm a foot and a half high for mulch. In the mounds, the team created micro-irrigation tunnels. Then they began planting, first hardy desert trees to perform multiple tasks such as shading the understory crops, slowing evaporation of water, returning nitrogen to the soil, and providing windbreak. Then they planted a row of fruit trees: figs, pomegranates, guavas, mulberries, and citrus. The result: Within four months, they were harvesting figs from only three-foot-high trees. The crops were flourishing. They asked experts from the local university to come test the soil to determine whether they were successfully growing crops in salty soil, or if somehow the salt content had diminished. Indeed, they found that the soil was becoming less salty. Not only that, but rich topsoil had rapidly accumulated.

Eventually, funding for the experimental project expired, and the site was left to local people to perpetuate. One might assume that the desert would take back this terrain. But amazingly, from last reports, the growth continues, because the system was set up to work within its environment, not to fight it with brute force.

This may sound minor, and it is obviously not at the scale to feed millions, but the principles are very important to what we can imagine for our towns and even cities. If an outdoor stretch of the deadest desert can be made fruitful with only rainwater—and no pesticides—then certainly we can imagine and intend for richer territories to produce abundance in similar fashion. As we will see, even barren rooftops.

The Greenhouse Effect (This Time It's Positive)

So we have a vote optimizing the earth battery for sustained charge, and for the best materials to power that charge. The other factor we can consider is optimized dimensions. Ideally, a battery occupies the smallest dimension while providing the greatest power.

When considering the earth battery, one might be surprised to learn that the Netherlands is essentially one of the most streamlined, minute organic batteries one could imagine. Surprisingly, that smaller country is second only to the United States in terms of agricultural exports (in financial terms) from its production of traditional crops, tomatoes, dairy, and flower bulbs. How in the world did such a small country do this? How could this be done by the second most densely populated country in Western Europe, about the size of Maryland or Bhutan, or slightly larger than Haiti?

The secret is that the Dutch manage nature and its forces very well in open farming. But they also use greenhouses on 0.25 percent of their land, which allows the country to be hyperproductive per square foot, eliminate wind damage to crops, increase solar flux, and reduce water evaporation; furthermore, the soil nutrients exist in closed systems, making their reuse simple. Not only do the greenhouses increase crop yields and decrease energy and water needs, they actually can generate heat for adjacent structures.

If the Netherlands can produce this much value on so little land, what if that country's methods were applied to other places? Greenhouse growing would allow us to reduce the transportation costs for food by producing crops closer to urban centers where they are needed. And we don't have to think only about the usual horizontal greenhouse. With a vertical greenhouse where planters are stacked, the rate of production per square foot of land can be as high as six times that of open farming in soil. Crops could be grown on, under, and in buildings to serve local markets.

Human beings naturally upcycle by migrating toward cities to live closer together in compact units. Urban density, by its nature, can deftly enable effective and efficient resource use while encouraging creative and diverse cultures of all kinds. Yet so much of the space in cities is underused. Certainly, we are good at packing as many people as possible into vertical space, but there is a territory in the city almost as large as the city itself that goes unemployed in the project of abundance: the rooftops. Cities and buildings, especially well-planned ones, can be reconceived as gardens. One can imagine a city from the air looking like a large garden divided into a multitude of plots.

Already, cities all over the world are being improved by green roofs. In 2001, Chicago mayor Richard M. Daley hired Bill's design firm to conceive a green roof on City Hall. This ultimately saved the building $5,000 per year in energy costs. More important, it inspired other green roofs. Chicago's building code is being altered to promote them. Walmart installed its first green roof in Chicago in 2008, and its commitment to green roofs elsewhere is growing.

A former naval yard building in Brooklyn is now the site for a large urban rooftop farm—100,000 square feet of greenhouse. It is estimated that it will be able to produce one million tons a year of lettuces, tomatoes, and herbs, all hydroponically (in water),[11] and will

---

In traditional soil farming, the key limiting factor is the active transportation of nutrients to the roots. Freshwater aquatic systems are ideal media for vegetation. Salt-water

sell the produce year-round to local supermarkets and high end eater-
ies. Bill's architecture firm is now designing schools, offices, and factories, which are covered with solar collectors and greenhouses, accruing
energy and producing organic food, clean water, and jobs.

Think about what could happen if we began utilizing all
available space this way. During World War II, victory gardens, home
plots planted with vegetables to help reduce the strain on domestic
food supply and on transportation of goods, increased vegetable production in the United States an estimated nine to ten million tons,
nearly equaling commercial vegetable production at the time. Those
were dispersed gardens, using available green space or creating
green space where none previously existed.

What if, in the same way, human beings upcycled places
one might not think of to provide earth battery power wherever
needed? Think of a simple strategic instruction: Where possible, go
from gray to green and hard to soft. From asphalt to vegetation,
from concrete to earth.

This is not just about reclaiming underused spaces in established cities. New cities are being built from the ground up in
tough terrain for farming in the surrounding land. Bill's firm has
been working with various Chinese cities that have problems with
flooding and populations predicted to double. He has proposed
bringing the "waste equals food" concept to life: building fertilizer factories, where struvite, magnesium, phosphorus, and nitrogen are cycled, and using this to restore biodiversity to the city's
parks and gardens, while simultaneously cleaning drinking water.
The most dramatic change proposes to lift the farming onto roofs,
thereby working around the threat of flooding by optimizing water

---

agriculture is also a possibility: Tomatoes are being grown on salinated farmland in
Saudi Arabia, for example

absorption, storage, and usefulness for park and food production. For a new city expanding into previously untouched land, the area might look the same as it used to from the air—green stretches of vegetation—except underneath a city has appeared. Let's take this greenhouse thinking even further.

Grown with the Wind

As we said at the start of this chapter, the first green revolution, which could be seen as the dawn of agriculture and green growing, was dramatically expanded by the Borlaug green revolution of the last decades in technique and nutrient management, applied minerals, and genetic modification. The conventional industry thinks that the next green revolution will be an extension of this last one, and we know, of course, that those efforts will certainly continue, since powerful factors in industry and economics are working that way.

But society might also be delightfully surprised. The next green revolution might not come only from that direction. It might well come from intensive local growing and locally optimized systems that benefit from shorter transportation distances, optimized water use, improved permaculture to replace chemical requirements, and multistory greenhouses.

Or even greenhouses that don't need the sun. Today, farming with LED-lit greenhouses is much more expensive than farming using natural solar light, but we are watching closely as the price drops dramatically due to mass production. If the LED lights are run on renewable power, the system becomes even more interesting. Dutch researchers have found ways to stack farming in warehouses in shelves one and a half meters apart.

China has a real challenge feeding its expanding population. Some regions of China are considered too windy for good farming; 61 percent of the country's desertification is caused by wind. Some data shows only 15 percent of China's land is considered arable, and so the pressure now is on the marginal land to produce the food needed.

In China, roofs might be employed, through stacked hydroponics and the rejuvenation of soil through use of the country's biological phosphate and biogas off-products. But a city can't feed its whole population by itself, nor can it produce the full varieties of foods people want. The supply must come from the surrounding countryside. If you extend this idea of a concentrated and optimized agriculture system by even miles into the surrounding land, it begs the design question: Why do people go hungry?

If you look at the whole problem, these remote populations need food, but it doesn't make economic or energy sense to ship produce from far away. What if these regions built greenhouses, some even underground? In that case, the windiness of the regions would be an asset, making the territory *more* arable.

In this case, the light would come not from the sun but from the wind-powered LEDs. The wind would be bringing calories—energy to be used by humans—to people through the food. And what if you took that idea of underground growing even further and were not just getting the crops you needed but were actually storing energy in plants to export to areas that needed energy (in this case caloric). One difficulty with wind power is that it vacillates due to the changes in wind strength throughout the day and night; utility operators struggle with what to do with excess energy during strong gusts, how to store that energy for calmer periods. An ideal system is to create ways to fruitfully dump excess energy and have it ready for another time.

So if we think of the windy areas of China, huge gusts could come through, and unlike most power stations that don't know what to do with 100 megawatts that show up at one or two in the morning, the system could place all that excess energy into greenhouses and growing facilities, protected from the gale but using the wind to grow plants to make the vegetables. These plants could grow on long racks in meter-high stacks, using only the spectral frequencies of the light that each vegetable needs.

A battery is just something that converts chemical energy into electrical energy. Here electrical energy is converted into chemical energy. Earth as a battery, plants as a capacitor. Instead of metal batteries that are expensive and toxic, how about food as a battery, storing energy for our beneficial present and future use?

Now that's an idea.

Let
Them
Eat
Caviar

Chapter 5

In Akira Kurosawa's film *Dersu Uzala*, a group of travelers sets out to survey a region of harsh, wintry Siberian forest, territory impossible for them to navigate on their own. They encounter Dersu, a wise nomadic tribesman who knows the area and agrees to serve as a guide. During a storm, Dersu leads the group to a hut in the forest. Dry wood has been laid there already, left behind by the last visitor. Because of the previous visitor's forethought, the travelers can light a fire and warm themselves in the devastating cold. When the group is ready to leave the hut, Dersu is astonished to discover they would do so without stocking a new supply of dry firewood for the next person seeking shelter. How could they think of doing that? Why would *anyone* do such a thing?

From Dersu's perspective, the travelers should leave the place as they found it—in fact, better than they found it: upcycled. In the case of the dry wood for the next visitor, arriving tomorrow or years away, they would leave one more stick than they found. Why would you deplete what you have enjoyed without leaving at least as much for those who come after? Don't the next travelers deserve the comfort and shelter that you enjoyed, and in fact allowed you to survive nature's extreme ordeals?

We think Dersu is on to something. If all of us were to design, build, and live like Dersu—improving the world rather than depleting it—humans could enhance the quality of life for everyone, far into the future. They wouldn't leave a legacy of scarcity and fear. They would upcycle.

As strongly as we know and have shown that enterprises adopting Cradle to Cradle principles can be economically fruitful, delightfully effective, and superbly efficient—removing the need for regulation, litigation, and the like—our thoughts never stray far from the most important simple issue: fairness.

It is not fair to future generations to leave this bountiful world depleted.

It is not fair to taint the healthy offerings the earth can provide.

It is not fair to poison mother's milk.

We often refer to Thomas Jefferson's letter to James Madison in 1789 in which he stated that a federal bond should be repaid within one generation of the debt, because "the earth belongs in usufruct to the living . . . No man can, by natural right, oblige the lands he occupied, or the persons who succeeded him in that occupation, to the payment of debts contracted by him. For if he could, he might, during his own life, eat up the usufruct of lands for several generations to come, and then the lands would belong to the dead, and not to the living."

That odd word, "usufruct," means the right to enjoy or benefit from property owned by someone else or in common ownership, as long as that property is returned undamaged. The concept of usufruct and stewardship are as old as, and central to, many religions.

We can always be thinking not only how we can design for a healthy present but also how we can return our common property, our common inheritance, as we found it—and, like Dersu, be generous enough to provide a little more, like the sun. We want to think of transferring value.

This sort of value transfer means to make goods literally "goods," pleasing things, to design products that foster abundance in the world. Right now what we call consumer goods can too often be "bads" in the long run.

By definition, a Cradle to Cradle design intentionally, from inception, not only adheres to the conventional design criteria—cost, aesthetics, and performance—but also puts values at the top. We talked about this concept in our last book as well. Typically, companies use a concept adopted from the UN of a triple bottom line—people, planet, and profit. After the economic concerns have been answered, they work their way down to see what profit can be found in the two other metrics.

But we know that if you put people, planet, and profit at the *triple top line*, good effects cascade down and outward. Equity, ecology, and revenue generation—not just leftover profits, whether incorporated into the design for a lamp or a living room or a city—spark innovation. Using the triple top line, we, as designers, might ask: Is this design good for biological systems? Does the design promote fairness, not harming the people who make the product or use the product?

As we said in *Cradle to Cradle*, the top values don't have to be just ecology and equity; they could be more playful. For example: How can I make something that is as much fun to reuse—to send into the next cycle—as it is to use? How can I make something that is beautiful?

For people around the world, tyranny represents an extreme form of ugliness. If humans design without values such as beauty in mind, they will cause what can be characterized as "intergenerational remote tyranny": determining that those in the future will suffer and/or have less opportunity, health, choice, and freedom because of our actions today.

Values-based design delivers benefit (or, at the very least, during a transitional period, seeks to inflict no harm) to all people and species. It celebrates them all. And it lets the earth belong to the living—now and in the future.

Would human industry be perceived as negative if the results were beneficial and benign? Of course not. A washing machine delivers a real benefit. The same goes for the computer, the car, and a long list of things many of us enjoy. Desire for these things is not decadent but a perfectly legitimate preference for more ease and enjoyment in life. Abundance—of us, of our products—is not the scourge: Society can accommodate and encourage even hundreds of thousands of products, from thousands of cultures, and even honor every one of 10 billion people predicted to be here later in this century.

Indeed, the number, in theory, is limited simply by physical and creative resourcefulness, the things around us, and our ability to work with them helpfully and productively. A Cradle to Cradle world enables people to enjoy abundance, to cater to diversity, without impoverishing the world.

Caviar Is Toast: What to Do?

Let's contemplate an irony: the scarcity of luxury.

Bill once had a Wall Street banker to dinner, and when he asked him what the secret of Wall Street was, the banker answered, "The creation of the perception of scarcity where none exists."

Recently, we came across a story about caviar and its shortage, which reminds us of this notion. Caviar, of course, is considered by many people a quintessential luxury item—costing up to $2,400 a pound. One might even think price could have protected the sturgeon population, since not everyone can afford such a food. But royalty, and then just the rich, and then the middle class, consumed caviar with such gusto that the population of various types of sturgeon, which provides the specific eggs, dwindled.

Trade restrictions have now been placed on the sale of beluga caviar from the Caspian Sea. The United States, after being one of the largest producers of caviar in the early 20th century, lists sturgeon as an endangered species. So caviar consumption might have become an experience of the past. Previous generations did not honor the usufruct of caviar, and it looked like no one would enjoy caviar in the foreseeable future.

Of course, this is a luxury item, not a basic necessity like water, but it brings to mind Garrett Hardin's essay "The Tragedy of the Commons," written in 1968 and still relevant.

Hardin was discussing the practice in medieval Europe of using commonly owned pastureland for grazing privately owned cows. In this unregulated system, he argued, an individual cow owner kept adding animals to his herd, even if it harmed the common good. If the cow owner had four animals grazing for free, why not five? Or six? Or more? The larger community of farmers, on the other hand, would suffer at these additions because too many grazing cows led to a deficit of green grass. The farmer might have increased his own personal benefit, but at the same time he (probably unintentionally) destroyed everyone else's potential benefit (including, later on, his own).

Hardin's idea of this predictable pattern is a profound argument for regulation around the world to keep our natural "resources" from being depleted at rapid rates—including other seafoods such as cod, tuna, and salmon, the rain forests, and so on. These regulations ask us to look at the commons—our common inheritance—in a way that supports and enriches us all.

Nowadays our commons might not be immediately obvious to us. The ramifications are so distant that it is as if you sent your cow out to graze in Vermont and a wetland drained in Indonesia.

An example: When Bill was a child, his grandparents lived on a gravel beach on Puget Sound, north of Seattle, Washington, where they harvested oysters. On summer visits, the family dined on oysters at almost every occasion. When guests were over for dinner, his grandparents prepared oyster roasts over driftwood campfires on the beach. They canned cooked oysters in glass jars and filled the freezer full of oysters in reused waxed-paper milk cartons for the off-seasons and to give to friends and family from the city. In other words, they perpetuated, harvested, preserved, and shared the oyster bounty in usufruct.

As Bill was growing up, his father would tell him and his siblings, "You have endless potential. The world is your oyster." That comes from Shakespeare, meaning a person could enjoy anything the world might offer with ease. The oyster might contain a pearl. It might provide delicious eating. It was free for the taking. The planet has endless potential—this one planet.

But ocean acidification has become a serious concern in recent years. The first real data points are coming in from an oyster-growing area in Oregon. They show a slight increase in ocean acidity due to the increase of carbon dioxide from the atmosphere. That slightly increased acidity inhibits larval growth for the oysters—the growth stage before and after the oyster creates its shell.

The world is no longer our oyster.

The oyster is like a canary in the mine shaft of ocean acidification. What does it mean if the oyster—our oyster, our potential—is dying out? How can humans fathom the scale of our actions if turning on a light in our home might lead through the cascade of effects—the use of coal-powered electricity, to the carbon emissions in the air, to the loss of oysters in the Pacific Northwest? This is the butterfly effect of chaos theory. People are starting to realize how these small actions can echo through larger systems. In the next chapter, we will talk about our version of the butterfly effect, how people can upcycle the butterfly effect to cause positive, abundant changes.

We sometimes worry that the human race has become too focused on how to simply use the resources of this planet. We even see terms like "natural capital" simply referring to nature for the service it provides. A chair is here for our use; therefore, the elements that make it are part of the usufruct. Some parts of the world might not always be here for our use. For example, the rain forest. Perhaps people don't have to tear it up for fiber, for wood. Perhaps humans don't have to see the rain forest as an air-cleansing system. Perhaps people don't have to see it as a giant oxygen-generating machine for our benefit. They can just think it's beautiful.

In our work, we like to consider what is the necessary relationship and what is the unnecessary relationship between humans and the rest of the world—or, we might say, among all elements of this vast biological system. We don't have definitive answers for this, but we think the question is an important one.

So how does this connect to caviar?

The Japanese enjoy some foods for their quality of umami, sometimes characterized as an ineffable deliciousness. Caviar may not seem like one of the most important necessities in human life. Certainly, given concern over all the people in the world going hungry, this may seem a frivolous interest. On the other hand, if eating caviar isn't harming other people or ecosystems, why shouldn't it be as available as other river food for those who are inclined to eat it? It is an expression of our diverse tastes and cultures. You might think caviar is the greatest food known to the spoon; someone else might despise it as frivolous or nonvegetarian. But why should someone who does not enjoy eating caviar be able to decide that you don't need it? Just because you don't enjoy chess doesn't mean it should be eliminated.

All these conundrums of diversity can be positively engaged through dialogue and intentional design. Diversity makes us more resilient and adaptable—able to evolve in rich ways—as a species. If the nearly seven billion of us all wore cork-soled sandals and cotton, the world would run out of cork and dry up. In that context, diversity might do us some good. In the same way, some people rejoice that some of us love caviar and some people can rejoice that love chess.

Caviar didn't have to disappear from people's list of possible pleasures. But clearly caviar harvesting needed to be reconceived. Here's what happened in one case.

A recent article in the *New York Times* described the Korean entrepreneur Han Sang-hun, who visited Russia more than a decade ago and saw firsthand the passion and market for caviar. He brought 200 sturgeon back to Korea on a private plane with the idea of cultivating a sturgeon farm to harvest the eggs.

But sturgeon are not ready to bear eggs early in their lives. Han had to care for the fish for 12 years, waiting for the moment they would be fully mature. In the meantime, he devised ways of extracting the eggs without harming the precious fish. Those fish, once the eggs were harvested, would mature for another two years until they were ready to be harvested again. This required tremendous patience and dedication on the part of Han. Despite the arduousness of the venture, Han now enjoys full financial satisfaction from his program (his caviar production accounts for 10 percent of global production). And the enterprise made him think about these fish in the larger context of the world and even the future days beyond his own lifetime. These fish will endure for 50 to 150 years (depending on their sex; females live much longer).

"The fish will live long after I am gone," Han said. "I am thinking about who's going to take care of them when I am no longer here. Raising sturgeon, I have learned a lot about time, human mortality, and environmental preservation."

What a delightful story! Han didn't internalize his passion for caviar and then despair at the sturgeon's decline. He carefully cultivated life's own tendency toward abundance—the desire life has to always upcycle. He was patient. He thought about how this cultivated abundance would perpetuate itself, for the usufruct of all caviar lovers, far into the future.

Upcycling our attitude involves changing some of the vocabulary we have used to describe work to be done—even words we ourselves have used. For example, one of our long-standing principles, and one that we outlined in our first book, is "respect diversity." Now we recognize that this sounds somewhat didactic. For two decades, we have been very concerned about diversity shrinking due to mass production, modernization, and homogenization of local cultures, the popularity of churning global fashions, and other factors. We still are concerned about these things. In the past, using the word "respect" hammered home our design assignment and voiced a rule for ourselves.

But you can respect the endangered golden lion tamarin in Brazil. And you will be paying your respects long after they are all dead.

If you do nothing to support them, encourage them, design for them, they will disappear. You can't just put them in a frame and respect them.

We want to ask you now to consider *celebrating* a delightfully diverse world—a much more positive term, indicating pleasure and enjoyment. A celebration is something people look forward to. If you celebrate someone or something, you're honoring him or her openly, without guilt or furtiveness. And usually you are having fun. A celebration also tends to be something that a group does together, commonly inspired. Let us agree on this as our destination—a world we celebrate.

If this planet is going to support billions more people than are here now, we want to look at each new person as a joy, a neighbor, a creative contributor to the common good, not as a burden. What might each person invent to improve our world? How can all of us support their ingenuity? What is fair for them? What can people design for their health and longevity? How can we express intergenerational generosity?

If we know that the water coming out of the textile factory is as clean, or even cleaner, as the water coming in, we know that our presence, our industry, our commerce, is good for people now and future generations. Human beings could be in this way designing for 10 billion people, since in areas with clean factory effluents, farms, and solar collectors on the roofs, it would not matter if there were 10 factories, or 150, or 5,000. Each factory would be a known good. We want more, not less, of them.

People can also design so that what they create does not impose itself on the tastes and needs of future generations. Bill designed the Bernheim Arboretum visitors center in Louisville, Kentucky, to literally come apart. Not now. Not for a good long while. Not until the visitors to this beautiful research forest want the structure gone. But when that moment comes, the building can be easily disassembled. The steel connectors and simple glass windows can be used again. The wood can be reused or returned to the soil.

This disassembly-and-return concept is now being applied to very large technical design projects, designing "What's next?" in things. We can create a bottle for drinking water, for safe water delivery, but we can also design for the moment that the bottle accidentally gets loose by the seaside and ends up in the ocean. We can design that bottle to be safe not only for drinking and delivering water but also safe as food for ocean organisms or as fuel if someone wants to burn it even in India—it can be designed to be safe. Currently, if you burn most PET (polyethylene terephthalate) bottles, you generate antimony trioxide in the smoke—a known carcinogen. We can design with myriad futures in mind, thinking how everything we make will move through the world, how it might eventually decompose, or how it will be used again. Over and over, before we even start, we can ask: What's next?

We think always of the upcycle: Optimize materials or their ingredients. Optimize product pathways. Optimize nutrient management. If these ideas sound opaque at the moment, just wait. We will explain.

Here Today, Where Tomorrow

People leave behind nutrients not only in the biosphere. They also express their intentions in the technosphere. There is much talk today about national debts, budget deficits, and raising deficit ceilings. One U.S. study notes that every baby born to American parents is born with approximately $45,000 in debt to his or her name, before even holding a job. (Other studies calculate the amount at three times as much.)

We think society is setting up a similar debt when it comes to technical nutrients. "Planned obsolescence" dates back to the 1930s as a mode of stimulating economies, by making consumer goods break down or go out of style after a particular time period so as to instigate more purchasing. It gained currency in the 1950s.

But because of society's design strategies, that planned obsolescence brings with it a significant debt in terms of the raw technical materials put out of industry's reach in landfills.

We know society has the capability of being more careful with its raw materials. How do people treat gold, for example? Because society values gold, no one simply throws it out to be mashed in a dump or melted into a monstrous mess in an incinerator. That is unthinkable. Everyone would wonder how anybody could accidentally throw their gold away and how we could dig that gold out of the mountainous tons of waste to reuse it.

Instead, people traditionally sell it in its whole form. Or they pass it down to their children. Or they sell it to be remelted and made into gold of equal value.

Now think of cobalt, used in medical implants. Indium used in LED lamps. Neodymium for wind turbines. Lithium for batteries.

These rare-earth and heavy metals are truly precious because they allow us to have the needed and valued goods, such as lifesaving devices, renewable power, computers, cars, and so on.

But if people keep designing for one material use and not reuse, we "use up" clean forms of the technical nutrients needed to make the products for the future. This means we will all worry about "limits to growth" because we feel we are running out of resources. Because of suboptimal design of virtually all current appliances from a material-reuse perspective, there's a chance that the technical nutrients used to make them are being used up. The same goes for computers and cars and lawn mowers.

Just as with fossil fuels, the quantity of metals and basic elements held by the earth seems vast. But ultimately these technical nutrients are limited. In truth, the only incoming recurrency available to humans is solar energy, rainwater, and the occasional meteor. In our work, we have started calling these metals "endangered technical species" to convey the seriousness of their potential loss to us in pure form: Extracting the metals requires lots of energy. If people reuse them, far less energy is required for their (potential) recapture and reconfiguration. Yet people do not recycle them as well as we could.

Like fossil fuels, the metals are capital being spent as if they were currency, and people are contaminating and depleting what could be available in pure form for generations.

Also, like fossil fuels, the technical nutrients are dispersed unequally around the globe. China currently supplies most of the world's rare-earth metals, and it has used that fact to express political displeasure toward certain trading partners from time to time, sporadically interrupting the supply. Nations need not be in a state of anxiety about access to productive materials when people can be using and reusing the resources already available more wisely.

The metals, of course, will never completely disappear. They will simply become less accessible and more expensive to extract. Our descendants might be mining copper not in large chunks but molecule by molecule in vast urban dumps. They will also be dealing with the toxic presence of these technical species in the water and air.

Years ago, Michael analyzed a television set to see how many chemicals it contained. The answer: 4,360. So many chemicals you the customer wouldn't even want to know, and those components were off-gassing into your home every day. With Michael's help, Philips produced the Econova television in 2010, which is designed for almost complete disassembly. It is PVC-free, and its cables are halogen-free. The TV uses only 40 watts of power when on and even has a solar remote control.

We know we can design better for the planet. We know we have the power to do so.

Such material misuse has grave repercussions for social fairness. It is certainly not fair for children to be born into a world with a deficit of pure technical nutrients.

It is not fair to leave a clean air and water deficit because our processes and manufacturing emit polluting elements of all kinds into the air and water around the world.

And it is not fair to leave a safety deficit because so many harmful and toxic chemicals are put into products with no idea of where they will end up or how to prevent them from leaching into biological systems.

It is certainly not optimal by any means for children to be born into a world in which unnecessary chemicals (meaning chemicals that do not optimize healthy human or organic life, such as flame retardants) are accumulating in natural systems. It is certainly not fair to design products and systems that cause irreversible damage. And it is not fair if mother's milk is contaminated by chemicals that are bioaccumulated through exposure to toxins and would not be legal to sell on a store shelf. We call that "chemical harassment."

There are more than 150 studies showing that Bisphenol A, which is used in the production of plastics, is harmful to our health as well as to the environment. Only industry-financed studies show BPA to be relatively harmless. Issues like this one are not about suing someone or about placing blame.

Start with good intentions right from the beginning of the design process. Optimizing materials means choosing the fabrics or metals or polymers that begin with goodness in mind.

Sometimes, when a technical nutrient known to be a toxin or endocrine disrupter exists in a product meant to be used in biological systems—BPA in a baby bottle, for example—it is replaced. But the substitute ingredient seems "less bad" only because people do not know enough about how the chemical interacts in biological systems (many of the substitutes for BPA use the same processing that causes the problem with BPA). This blind use is not a solution.

Some chemicals that are only suspected of being troublesome are kept in formulations even as the chemicals undergo decades of study. Their use around the world is prolonged, pushing risks into the future. We can ask, "Why still use them when we don't know how they will react in the body or other biological systems?"

Then there are the chemicals excluded by regulation but without solutions attached. For example, legislation called for the removal of lead, among five other elements, from consumer electronics in Europe, but the regulation stated only that the objects should be lead-free. It did not positively identify what *to* use. So here is what happened.

The lead solder was often replaced with a solder containing bismuth, which probably seemed like a good idea at the time because bismuth is thought to be less toxic, but it's toxic nonetheless when released to the biosphere; it still raises concerns as a heavy metal. Also, bismuth is almost never mined on its own, because it is not profitable enough. Usually it is extracted with tungsten or lead. So the resulting mining of bismuth, for use as a lead replacement, caused massive amounts of lead mining in China. The unintended consequence was more lead mining! Lead flooded the marketplace, costs fell, China's water quality deteriorated because of lead contamination due to the mining, and more lead ended up in other, unregulated products.

Is it fair for a standard or a regulation for a product in one country to cause damage and degradation elsewhere? Is it fair to respect the health needs in Europe while damaging the health of people in China?

Of course not.

Good design, with intention, with the goal of upcycling in mind, makes things better over time: just, fair, healthy, safe, quality for all—at all economic levels, in even distant places.

As we stated in 1992 in the second Hannover Principle, "Recognize interdependence. The elements of human design interact with and depend upon the natural world, with broad and diverse implications at every scale. Expand design considerations to recognizing even distant effects."

Fairness in design is not simply a moral matter but also one that defines quality. How "good" are you as a designer if the object you design causes harm, destroys the environment, or endangers children's health? Of course everyone can make mistakes or miscalculations. We all make them every day. They are inherently part of the creative life. But if you design knowingly using a toxin or a questionable material in your work, how talented are you really?

We can even look at this issue of quality in design using the old standards—cost, aesthetics, and performance. Is your electrical transformer insulation a good design if the cost of cleaning up the PCBs on the rivers where the factory is located or in the field where the equipment is deployed is astronomical? Is your child's toy aesthetically pleasing if she is in danger of brain damage from exposure to the lead? Are a soda and its bottle actually "performing" well if antimony is leached from the bottle into the acidic liquid and is dispensed to the person every time he takes a sip?

Quality in products and systems means they do not harm people, narrow their possibilities for life and liberty, or reduce their quality of life.

If people think of children, of the future, they can keep the focus on how to make products and systems intergenerationally generous. They can figure out how to generously give back while taking—and give back for generations to come. They can even start small in this thinking.

We have wondered why society seems compelled to only worry about the negative element leaching into our biological systems. The upcycle lets us think of moving beyond toxic and even beyond nontoxic to potentially beneficial. Bill recently designed a door handle that would transfer, if anything at all, magnesium molecules to your hand upon touch. Many people take mineral supplements in tablet form. Minerals essential to human health could be part of everyday interactions with functional objects.

Personally, we've found that in projecting forward while designing, with the greatest generosity in mind, we design for those

with the most modest means as well as for those who can afford anything they want, and for all generations.

Instead of the tragedy of the commons, let's upcycle
the commons.

Save Endangered Technical Species

So how could we change our actions to support the diverse desires of
people now and in future generations? We don't have to suppress our
desire for newness, for abundance.

Like Han, who created a system for the natural and effective
proliferation of the sturgeon, we can create a system for the natural
and effective proliferation of computers, of sneakers, of carpet . . . of
anything we want or need.

We can use the "product of service" concept and take-back
programs that we described in *Cradle to Cradle*—as basic as reclaiming whole products or as advanced as identifying and reusing particular chemicals. It makes sense for some products to be received
back whole—a Puma shoe, say, where the manufacturer would benefit from the supply of raw materials and the relationship between
producer and buyer would continue. Or Goodbaby, whose strollers
are designed with their next use cycle already taken into consideration; when the used strollers end up back in the factory, they can be
taken apart and made into something new.

We can make products that easily disassemble and with
parts that have a specified reusable content. This works very well for
products shipped between countries. A take-back program doesn't
make sense if a chair is made in southwest Michigan but is shipped
to an office in Mexico City. After 15 years, does it make sense to return the used chair to Michigan? No. Instead, if the chair is designed
the way Herman Miller and Steelcase design their chairs, the reusers

could disassemble the chair in 10 minutes with tools as simple as those found in the kitchen drawer and put the parts in the region's recycling system; even better, the recycler would disassemble the chair. The chair's technical nutrients—aluminum, steel, plastic, and so on—could go to a local or regional aluminum cooperative, steel cooperative, plastic cooperative, and be used again. Its biological nutrient fabric cover can go to methane capture and to make soil enhancers. The materials become part of Mexico City's wealth, its resources.

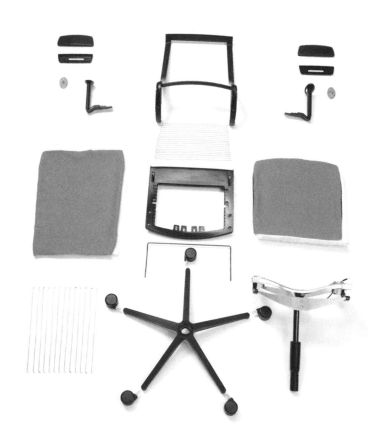

Upcycling Products of Service

Some of our work has involved what we called, in *Cradle to Cradle*, products of service, paired with a leasing concept. Obviously, leasing works great with many goods—cars, houses, land.

More than 25 years ago, we suggested expanding the idea to lease other goods as products of service, items such as large appliances, lighting, or carpeting. The manufacturer effectively retains ownership; a customer pays for the use of the product rather than the product itself, which can later be reused in the industry as a technical nutrient—food for new carpets, for example. When a customer finishes with the product, the manufacturer/vendor retrieves its technical nutrition; and the customer gets a new one, most likely from that very same manufacturer. The advantages are threefold: no "waste" of valuable technical nutrients for industry; the actual retrieval of materials by that industry; and a long and profitable relationship between a customer and a manufacturer.

The plan has worked extremely well for a window manufacturer that essentially leases its windows as products of service. When the company has a better window available, the customer can trade up, and the company reuses the old window materials. Solar services work well with this model too, in a relationship of constant improvement between company and customer.

Carpeting, it turns out, can benefit from a different approach to products of service. In the first attempts at implementing the idea, some carpet companies leased the carpet and retained ownership. But this arrangement left them technically liable for the carpet—if someone tripped on its edge and fell, the carpet company could be sued. It then had to pay insurance on the carpeting, which it had not had to do before.

An additional worry about maintenance arose; for example, some carpet-cleaning products degrade the fabric's quality or contain volatile compounds, which would make the returned materials less useful. These issues made the product of service and leasing concepts complicated as transactions for the manufacturer and its customers.

A more optimal solution exists, however. In Poland, for example, ownership is almost a religion, so the "product of service" is purchased outright, along with a deposit that is refundable when the product is returned "unharmed" to the manufacturer at the end of the use period. We have also developed a model where the manufacturer may choose to repurchase the product and its materials at an effective price in the future. The result is the same: The manufacturer gets back the useful technical nutrients, the product doesn't end up in a landfill, and the relationship between customer and manufacturer is maintained.

Our point is when preexisting systems and local needs are taken into account and sensitively interpreted, the product of service model becomes optimized. We might begin to characterize some products, such as chairs, for example, as products of disassembly whose component parts feed intelligent material pools.

Carpets serve as an instructive example, once again. The United States produces approximately 1.4 billion pounds of carpet "waste" every year. The carpet industry remains the last large-scale textile industry left in the nation. Why is that? Carpeting is heavy yet has low value per pound, so importing it overseas can be prohibitively expensive. With automated factories, its manufacture does not rely on intensive human labor.

What if, in America, 1.4 billion pounds of supposedly obsolete carpet materials could be continuously reused by the carpet industry? What if this liability could be recharacterized and processed as technical nutrients instead?

When the concept we refer to as "contingent assets" is implemented, the production of liability converts to the production of assets. Assets treated as currency—here today, gone tomorrow—become resources. And manufacturers, including a major carpet manufacturer that we advise, are beginning to do it.

Here is how it works: The manufacturer sells carpeting to the customer, delivers it, and installs it. The manufacturer has a deal that, for, say, 20¢ a square yard, it has the option to buy the carpeting and all of its assets back when a customer is finished with it. The benefits are twofold. First, the value of those assets is essentially set at the time of the sale; if the carpeting is made from petrochemicals and if 10 years after the carpeting is purchased petrochemicals have changed price upward (quite possible), then it is a valuable option for the manufacturer to buy back these technical nutrients and re-source them, rather than pay for new sources of petrochemicals. Second, the customer and the supplier maintain a relationship (the most valuable part of the transaction). The customer contacts the manufacturer when he is ready to get rid of the old carpeting. The manufacturer sends a truck to pick it up, and it is marvelously convenient and cost-effective for the customer to buy a new carpet from the same manufacturer because it can be delivered at the same time that the manufacturer picks up its technical nutrient assets.

In fact, a major carpet manufacturer we have worked with reports that 57 percent of its total sales come from Cradle to Cradle–certified products—and that includes 91 percent of its commercial carpets and 54 percent of its residential, along with rugs, sports turf, and hardwood, laminate, and ceramic tile floors. This company has just signed a contract to have the option of buying back its carpet from a hospital system for 20¢ a square yard. Depending on cost-effectiveness, it can buy it back or not. The long-term relationship with the customer remains the truly delightful prospect.

Similar tactics for upcycling products of service can apply to a range of goods that use technical materials (capital that is, unfortunately, being treated today typically as currency or downcycled to lower grades). We can now design computers and television sets in which all the parts are defined plastics and metals, glued together with a new reversible glue, so that when the product is heated in a disassembly procedure, the glue shrinks and the parts fall apart, greatly simplifying recovery of all the elements. This is one of the ideas behind the Econova that we described on page 157.

The upcycle is about what's best for everyone, every way, everywhere. What's next is what's next.

Intelligent Materials Pooling

To make the system truly work, however, manufacturers need to know what resources they have, where the resources are, and when they might be getting them back to use as nutrients in another product. Some materials currently have an easy path toward reuse—aluminum and paper, for example. But why not celebrate all the technical species? Companies might conceive of a system that tracks and plans reuse for all the technical nutrients in circulation in the world.

Nature manages the cycles of biological metabolism. Let's do the same for the technical metabolism.

We have proposed a clearinghouse for businesses to exchange information about what technical nutrients are available and even to help each other design, manufacture, and reuse the best materials. The result would be a mutually beneficial system between players along the supply chain. We call this an intelligent materials pool (IMP).

The principle behind the IMP is that technical materials—what we call technical nutrition—can be endlessly reused. If businesses collaborated to reclaim the high-quality raw technical nutrients from each other, they would be incentivized to use the highest-quality substances, since they would know they would be getting the pure raw materials back again, such as clear, food grade plastic bottles that go through recycling systems that keep them at their highest quality or improve them, returning them for use in the best possible condition, and certainly without downcycling them by amalgamating them into hybrid materials.

As with everything Cradle to Cradle, the benefit of the IMP would be more than just environmental. It would be economic. Valuable materials such as alloys, stabilizers, and polymers were designed for reuse; some polymers can be recycled more than 90 times without losing performance quality, while intelligently designed steel can be recycled endlessly. If they were kept in a clean system, there would be no need to mine new iron and other minerals, because the reused steel would remain at the same high original quality.

Intelligent materials pooling, a collaborative business-to-business management system for the technical metabolism—in short, a Materials Bank—would provide numerous benefits. If all products were encoded, the nutrient management companies of the IMP might, for example, ask retailers to please tell their suppliers to design packaging so that nutrient management could optimize value at the back end, sending that material on to a different manufacturer. Value could be monetized and exchanged. Designers would know the parameters of the packaging they were creating. Whatever was produced would be optimized for the multiple parties concerned.

Reading the Technical Ingredients for Sorting

Valuable technical materials can easily be encoded with information, what we call "materials passports," detailing all of a product's ingredients and properties, so that when the product moves from one process or industry—and one country—to another, its makeup and technical nutrition is known and communicated. This can be done with simple molecular markers or infrared signatures.

PET (polyethylene terephthalate) bottles are recycled in a stream with other plastics and often downgraded to another use, such as fleece for a jacket. The jacket is dyed, perhaps using colorants with heavy metals, and a metal zipper is added. That fleece jacket can never cycle back up into food-grade PET. The food-grade quality level of the PET is lost in the process.

But a large-scale bottling recycler in California, working with the major beverage companies, has executed a sorting mechanism that can read whether a bottle is clear PET, green PET, or another plastic, even including PLA-based plastic made from plant materials that are compostable.[12] With this sorting and separating system, the company can recycle 75 million conventional pounds of food-grade PET a year. It's a food-grade bottle-to-bottle system, not a bottle-to-speed-bump or bottle-to-contaminated-clothing system.

Techniques exist right now to enable marking of a chemical so you can find it later. Some companies doing careful chemistry might want to mark their chemicals so they can prove, for example, that their processed fluids are not the ones that escape and poison the groundwater. Some companies may not wish to have their chemicals known. This will be a very interesting area for society's consideration, because information is power. Insightful businesspeople will want to participate in and benefit from this new information age as it relates to materials and to society's desire and need for transparency.

There is a place for middle management in the handling of these technical species, much as there will be in the reclamation of phosphate. A potentially huge part of our future economy is just now coming into existence: the part responsible for that second cradle—the return to the cradle, where the new reuse periods begin (don't think closed loop). The current prevailing philosophy about products, where there's a philosophy at all, has been "take, make, waste" or "bury or burn." Perhaps it's even time to rethink "reduce, reuse, recycle" and replace it with "redesign, renew, and regenerate."

---

12. Not to be confused with the "plant bottle" recently introduced, which has been described as containing "up to 30 percent plant-based" plastic. This means that "up to" 30 percent (one might wonder if it could be as low as 1 percent) may be produced from sugars converted into the basic glycol that is an intermediary chemical, typically derived from fossil fuels in the making of PET. The bottle is essentially PET with a different source for part of its basic chemical infrastructure. And the bottle is not biodegradable.

Those two steps—aggregate and return—are beginning to be monetized in our mass-production society. "Waste handlers" are becoming "nutrient managers." Our hope is that someday they will even become upcyclers.

The city of San Francisco recovers more than three-quarters of all stuff thrown away, compared to the national statistic of a little less than one-third recovered. Their rate is so superior for numerous reasons: The city encouraged recycling by passing an ordinance; it partnered with what we would call their nutrient manager, Recology, which created the infrastructure and understood the wealth in the "nutrients" the citizens were throwing "away." It intensely educated its citizens about the how and why of nutrient management. The city is well on its way toward its goal of fully reusing its nutrients—what used to be considered garbage—by 2020.

Some materials have shown their value on the streets; bundled cardboard left on the sidewalk is poached from the recyclers by unlicensed haulers.

This nutrient management can be big business. The Van Gansewinkel Groep, based out of Benelux, has gone from being a waste management company to a nutrient management company. It collects and processes "waste" into raw materials and energy. The company employs nearly 7,500 people and has annual turnover of €1.2 billion.

Could eBay-like Internet auction sites one day be intelligent materials pooling brokers? Social media holds the potential for anyone to become the nutrient manager for his or her home, block, neighborhood, or city. The original recyclers were ragpickers who gathered the fabric used for the paper for the most valuable artwork, archival papers used today—rag paper. An upcycle if there ever was one: rags to riches.

What Can One Ragpicker Do?

About one-third of the average dump is filled with packaging material, nearly 65 percent of household trash.

Just as humans think of designing necessary products to retain, restore, and enrich on a continuous basis, they can and need to think about upcycling the systems they employ every day to make our world run. Packaging is an excellent example of something that industries, businesses, and people engage with regularly but that exists in a far from optimal design.

A great deal of packaging falls into the "big tent, small circus" category—a tiny product encased in a thick plastic-and-paper package some 10 to 20 times larger than the product itself. Look at a little flash drive—it's tiny but sits in a small billboard of packaging so it can be visible on the shelf. The packaging typically finds its way to the garbage (it would be nice if it went into the recycling bin; usually it doesn't, and even if it did, it is an odd amalgam of paper, plastic, glues, and inks).

Not only is the detritus of packaging considerable, it adds to your consumer costs—in cosmetics and toiletries, for example, the packaging can be 30 percent of the cost of making the item. Plus, there are deeper costs: the cost to dump the packaging or to burn it. Perhaps if the package goes to an incinerator, creating electric power, some value is returned as energy, but the conversion from a very particular arrangement of molecules into a less compelling form is a significant loss in value. Pollution has been created. Assets have been turned into liabilities.

Until there's an alternative that promotes the seller's interest equally well, people will have an abundance of unnecessary packaging. Eco-efficiency people would advocate making smaller "right-size" packages. There's the inherent contradiction that the seller wants you to notice its great new product, and if its $35 computer chips had to be displayed in a bucket, loose in bulk, where the chips could be sto len and would appear not special, retailers would balk and customers would wonder why such a "common" commodity was so costly.

How can we think of a solution that incorporates everyone's interests—the seller, the buyer, the store, and the environment?

As shoppers, we already are accustomed to baggers who stow our purchases for us to carry away. Often we see a jewel case for small valuable things—CDs were the first big example—that is taken off by the clerk. And many stores have security tags on things like apparel that must be removed or deactivated. We could upcycle these customary clerk-customer relationships to create even more value for the store, the community, and the customer. Clerks could remove packaging materials at checkout—or large retailers could employ specialists, the inverse of greeters. They could be the "goodbyers" for the good buyers who remove and consolidate now valuable packaging at the store's exit—so the customer wouldn't have to deal with all of it at home. The plastic or paper would become a materials source—a profit source—for whoever handles it. In certain cases it could be returned to the manufacturer, like palettes, for example, thus bringing down costs—and cost savings could be shared with retailers and their customers.

Such a solution hinges on the packaging being either biological or technical nutrition, easily segregated into materials pools and coherently managed or easily disassembled. In this scenario, the maker or seller can have as mammoth a showcase on the shelf as it wants, without a real economic or ecological downside. Packaging is transformed into something briefly used and then recirculated, and whose value is maintained at a high level. Some companies are already working toward this: Aveda, the cosmetics and toiletries maker, announced its own packaging take-back program, saying the point of the program is "to hold ourselves accountable for the products we make." Samsung has a similar effort with a similarly stated goal. Kiehl's, the skin and hair care company, recently restarted its program of taking back its empty product containers and paying the customer with credits toward free products.

What a wonderful notion—these companies are looking for ways to get back their materials and eliminate the concept of waste. When a company plans to receive the bottle or box or computer back, it will make the bottle or box or computer with materials that the company can use again.

We could also consider getting rid of much of our packaging altogether. Great products might be promoted and celebrated in new ways. The "billboard effect"—advertising products in their excessive packaging—could be manifested in a different form. How? The challenge is to reach the customer without having to wrap and brand each item.

Future store shelves and dispensers could provide glorious frontage, with detailed descriptions of the product's qualities. Instead of trying to tell the inspiring story of the product on each individual package, it would be done on the vertical real estate in the store. The customer would walk up to a display, pick up the product itself, and that's it. The product would go into the shopping cart. In the end, the product goes home; the unnecessary packaging doesn't. The customer came and got only what he wanted or needed but not all the other things he didn't.

Celebrate diversity. We all want different things. We all enjoy different goods. The solutions we devise can understand that reality and encourage it.

City of Dreams: The Future Helping the Past Improve

To take this idea to a larger scale, let's talk about even more tenacious problems. Let's look at designing for urban diversity. Often the celebration of diversity meets its biggest challenge when populations feel they are being encroached upon by newcomers. The diminishing store of resources makes people nervous. How will we all find jobs? Will rent costs soar out of reach? Or even more dire fears: How will we all get water? Or power? Or food?

In our work, we are often asked to consult with cities about their growth issues. We have learned how much potential strength and innovation can be gleaned by flipping this fear around. What if instead of thinking how an increase in population might drain a city of its resources, we think of how a population bloom would enrich the city, how it could make the city more abundant, wealthier in ideas—what Emerson might have referred to as "a calculated profusion."

One of the world's major capitals, Beijing, is projected to double in population size within the next half-century. Bill was asked to think about how elements of that expansion could be addressed in Cradle to Cradle thinking, or "the circular economy,"[13] as it is called by the Chinese. Here is one idea: mining the old city.

If Beijing were to go from 20-plus million people to 40 million people, one would have to ask where the new sources of water, energy, and food would come from.

Beijing is already trying to deal with the fact that water has become a desperately critical resource—the city has exhausted its two reservoirs and is now using up its groundwater. The water table has dropped precipitously. And there's discussion about plans to

---

13. In the 1990s, when *Cradle to Cradle* was first published in China, the government officials and universities that translated and published the text subtitled it *The Circular Economy*. Even now, people are using the term "circular economy" with Cradle to Cradle as its basis. We are honored by that use, but Cradle to Cradle is not a circle—not simply closed loops— it's a spiral that celebrates growth because of the employment of renewable energy.

redirect rivers from the south—whole rivers—projects far larger in magnitude than the Three Gorges Dam.

You have to think: What would be the most fruitful design for such a rapidly expanding city, one that can be beneficial to its inhabitants now and in the future? Is digging up more streets, building more dams, producing more coal-fired plants the answer? If our goal is intergenerational generosity and a celebration of the commons, then the answer is no.

Instead of trying to scale the energy and water demands for the city doubling as almost a clone of itself now, we might think about how to work with what resources we have and design the new part of the city to meet those needs.

It is not uncommon for real estate developers to be asked to provide public benefits in exchange for the permission to develop building projects in sought-after locations. A developer might be asked, for example, to build a certain amount of low-income housing in exchange for the rights to build luxury housing in a certain area. In the United States, a developer might win community approval only if he agrees to construct a park or provide a public arcade.

Going back to Beijing, the new city builders could be asked to mine the old city for energy to build its new structures. For example, a developer wanting to build 400 new apartments would receive a building permit if he or she also took on the energy systems of buildings in the older city and upgraded them to help offset the needed new demand and also to make a set energy contribution to the grid. Let's say an inefficient old building uses 200 units each of energy and water per capita. A new building using current technology for construction materials and energy and water consumption can be easily accomplished at 50 units of demand.

The developer of new buildings would reduce the consumption of an old building by half, or provide new renewable energy and water to the mix, in order to get the permission and resources to build the new structure serving as many new people. The upshot would be more growth plus a revitalized city that can last, and flourish, far into the future.

Think about it: new buildings at 50 units of water and energy per capita, old buildings going from 200 to 100 units per capita. A city doubling in population ends up using 25 percent less resources with twice the people. Where did the water and energy come from? The city has overall reduced its energy demand by 25 percent. No new dams, no new power plants needed. The city has mined its past for its future.

Similar initiatives can be manifested to bring on additional renewable power. Often these are part of a corporate goal of moving toward certifiable renewable power in a cost-effective way. This idea may sound hard to imagine, but it's actually happening. For example, in our work with an international consumer products company, we recommended wind power. To date, 15 megawatts of renewable wind power have been brought online for their newest factory.

Beijing has worked to address its coal-fired power plant problems, in response to the increased air pollution. The city has been exploring the potential use of geothermal, solar (despite the smog), wind, and biofuel. As we said earlier, in order to explore the potentials, one might *not* want to start small. We can adjust our plans to practical reality when we need to. It's not hard to see the potential abundance in a bigger well-designed city.

An exciting project that might be employed around Beijing involves what we call "horizontal chimneys." Basically, as the system works now, as coal-burning electrical plants produce energy, they create CO_2 and other emissions, which plume into the atmosphere, the carbon out of reach for our use and the gases contributing to climate change.

But what if, when coal is burned for power, we could design smokestacks to capture carbon emissions and toxins? Instead of shooting the emissions into the atmosphere, where the carbon's value is lost and the air grows filthier, we contain them. We hold on to them.

The Ecological Sequestration Trust in London is funding the creation of substations next to power plants. Instead of the gases going up and out the smokestack, the gases will be funneled into algae greenhouses. Due to the high CO_2 emissions, the algae grow, rapidly, and when the tanks are full, the algae are sent through pipes to another station, where they are anaerobically digested to make biofuel and used for soil enrichment. Horizontal, not vertical, chimneys will sequester carbon for further use—retaining the asset—while also keeping it from making the air toxic—eliminating the liability.

There have been many experiments in this area, including advanced work at MIT and other major universities, resulting in various degrees of effectiveness and success and potential for commercialization. The question is a simple one: Just because chimneys have always been vertical, does it mean they *must* be vertical? Obviously, we don't think so. We can lay them down and put them to good earthbound effect—more than simply pumping the carbon from a power plant underground to be locked into a rock formation. Our vision is to put the carbon into a photosynthetic process where it could support plant growth. We are especially interested in algae because it can be as much as 20 times more productive than open-land farming.

This is the beginning of an innovation that might well lead to whole system optimization. (The tree in the earlier chapter cascaded from a piece of high-quality wood furniture, to particle-board, to paper, back to atmospheric carbon for soil enrichment, back to tree, with many human jobs created along the way.) In the same way, we can imagine a similar cycle of cascading benefit from burning coal. Conventionally, coal, a fossil fuel, is stripped via hydrogen combustion for energy production, and then carbon is stripped for more fuel production.

Instead, we see the opportunity for carbon sequestration systems to become economically interesting, especially when you add protein production. In the long term algae production is terrific for the generation of oils, cosmetics, pharmaceuticals, soil amendments—and fish food. Tilapia is the best example. Tilapia feed on algae; they are easy to raise on fish farms in ponds, and reproduce quickly. As a consequence, tilapia are inexpensive. The systems show the potential for profit because of the market desire for protein. Once we see all of these things combining into an optimized system, the benefits could become dramatic.

If Beijing and its surrounding areas were optimized around solar and wind potential because of its annual hours of sunlight—one already sees solar-heated water on most of its buildings—then the potential would be impressive. More solar collectors on buildings and parking lots. Wind turbines positioned as wind rows in the western plains to take advantage of the gusts that otherwise carry heavy dust into the city would create energy, and ways would be found to get the power to the city. People consuming goods could simply mean material for biofuel. More effective nutrient management replacing the sewage treatment plants of such a large city would mean more profitable and fruitful production of struvite for the nation's farmers. The more growth, the more good. The demand side and the supply side woven together in a beautiful tapestry of optimized growth. A city twice as big and living within its means.

Celebrating the Diversity of Solutions

In some instances, low-tech is the best way to meet a local need. Smart design is not necessarily high-tech. The woman in India who washes her clothes in the Ganges and can afford only one rupee's worth of soap a day can resupply the nutrient cycle if her safe soap is packaged as a small sachet that is a nitrogen-enriched biological-nutrient package instead of plastic trash creating annoying litter and clogging drains. People could collect and sell the nutritious packaging to the farmers for their crops, to increase soil health.

In fact, it has been posited that the part of the Indian population derogatorily termed ragpickers, who scavenge the dumps and the road litter for useful "garbage," actually performs a serious good for the country. They put 100,000 tons of "garbage" into the cycle of reuse every single day. Their economic value to India is more than $280 million a year, and their work feeds them and their families. It keeps the country that much cleaner. It saves all that energy to produce more glass and plastic and tin and everything else. This bolsters our point that if these materials were designed in the first place to enter the soil safely or as safe technical nutrition, that would create more incentives, better health, more materials in flow, the possibility of upcycling.

When Bill was a child living in Japan, his parents had a beautiful garden, tended by a careful, attentive gardener called Oji-san ("uncle"). In the garden of this old shogun summer colony, Oji-san sat cross-legged on a small grass lawn and, with his one small sickle, cut the area of grass before him. When he was done with that one small area, he would push down with his arms, his knuckles against the ground, and lift his whole body with his legs crossed, pivoting over six inches to the next spot and cutting the grass there. Then he would move to the next spot. During the growing season, Oji-san moved quietly and deliberately like this, cutting the grass until he was done. Then he would move on to his circuits of tending to the garden's needs, clearing the ponds, raking the gravel, attending to the trees. Then it was time to cut the grass again. Over and over, a lifetime of care.

One day Bill's father came home with a push mower, thinking that the new device would make Oji-san happy, saving him time and energy. But Bill saw Oji-san's face fall. He was stunned. He had no idea what to do. You see, the silence and the careful, constant tending are part of what makes Japanese gardens authentic. Efficient, mechanical, quick mowing would change all of that and make Oji-san less valuable and necessary. Oji-san had committed his life to tending this garden, to working in silence, and to clipping grass by hand. He practiced what the Japanese call *shokunin*—a social obligation, physical, mental, and spiritual—to care for this beautiful place. He and the garden were one organism. After seeing Oji-san's reaction to the new device, Bill's father quickly took the mower back to the store.

The point of the story isn't that people should never use lawn mowers or other mechanical conveniences, but that there are all kinds of ways to celebrate diversity, and some might not necessarily be efficient. For that particular garden, for Oji-san, at that particular moment in time, the traditional Japanese way was the best way. In other situations, a mower would work. Oji-san offers the ultimate example of tending a garden, of intending a garden, of a graceful and timeless mindfulness confronted by modern life's potential for timeful mindlessness.

When we celebrate diversity, we can say that we want the caviar lovers to have what they like, as long as they enjoy the caviar in usufruct. We can also thank the ingenious solutions of this diverse planet for helping us flourish in perpetuity—with endless resourcefulness of people and things and infinite care.

The path to a more beautiful world can come from vast plans and small gestures, it can save cities or tend sturgeon, as long as the strategy bears the needs of future visitors in mind. As long as one continuously rebegins every day with generous intention, as long as one honors the child frolicking in the garden, watched by the "uncle," his sickle fallen still, enjoying the sight of the boy amazed and wondering at the erratic freedom in the journey of a butterfly.

The
Butterfly
Effect

Chapter 6

You may have noticed that we have mentioned butterflies a few times in this book. The optics of butterfly colors. How MacCready, when looking for inspiration while designing his human-powered airplane, remembered the butterflies he watched as a young boy. The butterfly effect of our actions, how flipping on the switch that asks for a surge of carbon-burning electricity can create a cascade of reactions leading to the acidification of oysters.

We have talked a lot about butterflies because butterflies are nearly useless.

That's an exaggeration, of course. But it's sort of true. What serious function does a butterfly have? What economic value? A little pollination here and there for what? Milkweed? And yet, how extraordinary is it that child and adult alike note nearly every arrival of a butterfly on a plant or its path floating across a yard. The Annenberg Foundation in Los Angeles recently created a free app to track the monarch butterfly's migration. More than 900,000 schoolchildren and adults have signed on to contribute data, to track the journey! What a delightful game to play!

There is no more delightfully serious function in life and in business than to create joy.

That's why, when asked to create a pharmaceutical laboratory in Barcelona, we decided to celebrate the butterfly through intention. The laboratory would be both a highly functional workspace and a cocoon. While the project waits for the reemergence of the Spanish economy, its design is something we consider a delightful evocation of our upcycle work.

Let's look at the design itself. The building has many of the usual elements one might craft to make a structure eco-effective. Every façade is different, based on its orientation. Southeast is full of sunshades; southwest is solar panels. Northwest, on the lab side, has shades for the hot low afternoon sun, shielding the laboratories, and these shades come in hues inspired by the coloration of butterfly wings. On the north side, vertical plantings create a dust-collecting, oxygen-producing "green wall," using the building's gray water—the lightly used water drained from sinks and such, filtered through sand—to irrigate the plants. The number of plantings is precisely sized for the amount of gray water that the building produces.

The inspiration for the building design began from the floor up. For an earlier version of the design, we conceived of two round-cornered triangles joined by an atrium. When we began planning the floor-tiling patterns for those spaces, we found an opportunity—to save money. If the tiles had been mathematically cut to fan out equally, it would have been a very expensive proposition.

Bill was looking at the plans and said, "These look like butterfly wings. What if the tiles went down first and the walls got laid on top, and what if the patterns were the big beautiful designs of butterfly wings?" An economic problem turned into a delightful opportunity to save money and solve the problem with creativity and connectivity to other ideas.

We could echo the coloration of actual endangered butterflies in Catalonia. Anyone walking across the floor from one laboratory to another would be reminded of these amazing creatures that live in their parks and flowerscapes but now need support, need celebration. Each floor's tile pattern would be a different butterfly's wing.

We looked at the patterns of the wings of these endangered creatures and picked the ones we knew we could make right away using clay tiles and safe colorants manufactured by local industry. Then we looked at all the butterfly colors not yet available in a safe formula and began working with manufacturers to achieve those colors too, safe colors, butterfly colors. That was an upcycle of what had existed before, simply based on the fact that we decided to design around butterflies.

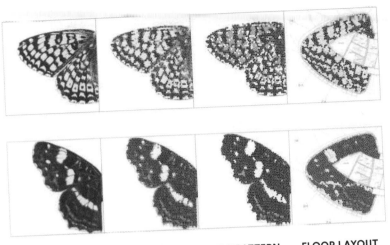

WING SAMPLE   **RENDERING**   **TILE PATTERN**   **FLOOR LAYOUT**

An exciting further evolution of the idea is our plan for the building's lobby. Using a usual budget for basic decoration, we thought, *Why not, right there in the lobby, run an actual, active butterfly hatchery?* We could partner with the local zoo to make this happen.

When you walk into the lobby, you'll see a glass wall to your right, an expanse of garden visible beyond the sheen. If you look closer, that glass wall is really two glass walls with space between and white shelves running across at regular intervals. Hanging from those shelves are butterfly chrysalids. Alive, changing before your eyes.

You might stop and marvel a moment at the metamorphoses unfolding in your presence, greeting you. You might even decide to return on the weekend, when the children and adults in Barcelona will be invited to release the butterflies and watch them fly off into the city. The outer panel of glass is configured as a door that allows the butterfly managers to tend the butterflies, gather them as they hatch, and finally let them go, in full wing, to the outdoors.

Once a week, this community celebration would occur, and afterward, the children might go home to tell their families about the experience—how the butterflies opened their wings, what it was like to see them lofting about for the first time. It's a beautiful process to witness, as what was contained and dormant reaches exquisite fulfillment. The children might share with their parents the names of the other people who were there to see the butterflies—students, the retired people who volunteer to tend the chrysalids, other children they have just met who have now become their friends.

After years of these chrysalids being nurtured, hatched, and released at regular intervals, these endangered species might even grow to a healthier number. Gardeners and home owners in the town might begin cultivating the flowers and plants desired by the butterflies and provide appropriate, even attractive, habitat. Highway verges once monocultures or barren could be revisualized as beautiful butterfly habitats, even seeded to encourage more butterfly activity. Green space in the city might grow. A new park could be dedicated to supporting the blossoming diversity. Bird species and other species would flourish too because of the returning native habitats and juicy caterpillars to dine on.

At the local school, the children, having had their curiosity awakened, might ask to learn more about species in the area, including other endangered species. They could become newly connected to their own place as they learn about its natural history and biology. They might want to know more about what they can do to cultivate and nurture these and other species around the world. The biology of the area would diversify. Learning and connectiveness—with nature, with the community—could deepen. Friendships would be fostered, and beauty that is far more than aesthetic would unfold.

Just think: None of this would happen if someone had simply used the same budget for nonnative indoor plants or artificial décor in the lobby for ease, convenience, and cost. This hatchery offers so much more: It upcycles the building, the community, the ecosystem. Useless. Fun. Joyful.

That is a design version of the butterfly effect. One small thing you do, and look how it proliferates when you let it go. Look at the abundance. Remember the child chasing the butterfly around the garden.

Many of the ideas we have discussed in this book resonate with us because of their tuning with a human being's innate sense of fairness, generous behavior, the sensation of working within the natural tendency of life to upcycle.

Not just individuals but businesses too can think of themselves within this tuning. Businesses could imagine themselves as positively minded NGOs (they are, after all, nongovernmental organizations), which set out to make the world a better place, to provide great value to the world. Most businesses today, if they want to do good works, do them "on the side" as corporate philanthropy. This can, of course, be a marvelous benefit to their selected beneficiaries. But in the upcycle, a business in itself can actually do positive work at every moment, in its services and in its products.

Take the hedge fund executive who spends his week on Wall Street, engaged in trading abstract derivatives using various financial instruments, many of which may be connected to immensely destructive behaviors, such as the decimation of rain forests or habitats, and then, on the weekends, heads up a national bird-protection organization. But what if that person spent the week doing business that actually made the world better for everyone, including the birds?

Inspired business leaders together possess the collective power to reverse our environmental woes in a way governments cannot. The best executives have profitable short- and long-term goals and profitable ways to get there. The government, meanwhile, in many countries, may suffer from corruption or not be constructed or sufficiently motivated to focus on long-term fiscal health—its or anyone's. When a business determines that its strategies have become tragic, it moves quickly to strategies of change or it dies commercially.

Numerous benefits accrue from thinking of business as being integrated with love of people (the definition of "philanthropy," after all). It clarifies and broadens, in the best way, how a company views itself. It changes how others view it: The products the company sells, its way of doing things, indeed its very existence, can be living testimony to its support for a world of prosperity, social equity, and environmental health.

Finally, a company conducting business this way does not always need to cede the moral high ground to cause-related NGOs or activists, who often position themselves as ethically superior to business. This dynamic has at times spurred the perception that environmental organizations exaggerate dangers and scare people, because it appears to be in their best interest for fund-raising. But aren't we tired of companies having to be regulated all the time? If I'm the head of a tire company, which is a faster and better driving force (pardon the pun) for positive change—the company's internal critique of an airborne-particle problem or an activist's public accusation? Wouldn't I want to improve tire quality so it's not in any way producing respiratory problems?

Cynics may say that we are being hopelessly naïve, that a business will do everything *not* to solve its problem, but we're more hopeful than that. While these are still the early days, there is something to see here. Just look at the proliferation of organic products, which are now sold by such huge mainstream retailers, where they weren't available a decade ago. When the retailers noted the success-ful upward trend of demand for organic products, competitors started to move to fill the market space. Although just becoming mainstream, organic food globally is now more than a $52 billion industry.

Readers of our first book may remember the design Bill's firm did for the Ford Motor Company's legendary plant at River Rouge. The "living" roof on the auto plant provided habitat, decreased the building's energy costs, and protected the roof membrane from ther-mal shock and ultraviolet degradation, thereby extending its life. But that's not all: It also acted as an on-site treatment plant for cleaning storm water through natural filters, thus exceeding the performance called for under the Clean Water Act.

The living roof garnered attention for its ecological sensi-tivity—a green roof with nesting birds atop an automobile factory!—but it was also recognized as an on-site water-treatment plant that resulted in an up-front capital expense savings of $35 million ($13 million for this natural storm-water-treatment biological plant rather than the $48 million high-maintenance chemical treatment plants with endless buried pipes that had been budgeted to be installed). With car sales at a 4 percent margin at that time, this $35 million in capital expense savings was effectively equivalent to getting an order for $900 million worth of cars. When the board saw the first cost capi-tal expense reduction, plus future savings, they approved the design in a matter of minutes.

The most effective transformational foundation of Cradle to Cradle is, to the surprise of some, not environmental. Nor is it ethical. It is economic. If Cradle to Cradle fails as a business concept and in-novation engine, then it fails, period. It succeeds when it celebrates economic growth, which in turn grows ecological and social revenue. It succeeds when it upcycles the economy, and ecological and ethical benefits accrue.

Creating Abundance at the Molecular Level

There is a butterfly effect every time a company signs on to the work of Cradle to Cradle. Not only does it strive to make its production more profitable and healthful, the knowledge it and we gain from the process causes tremendous ripple effects across industries.

We are always pleased when the poetry and big-picture perspective of Cradle to Cradle inspire people to build the upcycle into their organization's mission, operations, and procurement. One major company we work with changed its Corporate Social Responsibility and Environmental Reporting Program from "Total Impact" (TI) to "Positive Impact" (PI) to inspire its people to strategize toward beneficial goals, not just measure and aggregate data. Of course, much of the most crucial work, though, is then achieved in the lab, in push-pull discussions in the office, in conversations with suppliers. It's all about looking at what you have, analyzing it, and figuring out how to make it better. (Or, as we like to say, inventory, assessment, optimization.)

We have said before in this book that moving toward creating products that can flow within nutrient systems cleanly in a Cradle to Cradle way has many benefits: beauty, health, fairness, and profitability. But it is not easy and requires dedication and determination. Especially for those companies with the courage, vision, and wisdom to be in there and go first, to be the progenitors.

Just to give you an example, to redesign conventional paper towels so that the chemicals used are safe for biological systems, you need to replace 29 or 30 chemicals. That's just paper towels.

Now think of that television set: 4,360 chemicals. You understand just how dedicated a company has to be to solve the challenge of getting to Cradle to Cradle design.

But we have found with so many executives, marketers, managers, and industrial engineers and chemists—who discover inspiration and integrity in the process—that the effort is worth it a million times over. The beautiful outcome can serve as a seed for change across all sorts of industries. When we redesign one product with intentionality, a whole world of other related products has the possibility of becoming safer, healthier, more effective. For example, if a washing-machine manufacturer wants to rethink its designs, it could consider the Cradle to Cradle approach to materials, leasing, energy quality, water quality, and social fairness in textiles and electronics (and even in paper, for its instruction manuals).

This sort of cascading effect occurred when we were working with the United States Postal Service. The USPS was already "efficient," in business terms, when Anita Bizzotto, chief marketing officer, began to learn about Cradle to Cradle notions of technical and biological nutrition. The USPS had worked to create the lightest packaging for the needs of its customers. It moves massive amounts of products around the country every day. That is what you call business efficiency.

But looking to make the system more eco-effective, Bizzotto saw the opportunity for the USPS to be an enormous "food" source. When she traveled to MBDC headquarters to understand the implications of Cradle to Cradle, the USPS had already achieved the usual benchmarks of "sustainability." It had complied with EPA guidelines, both mandatory and optional.

Bizzotto was seeking help to transform the agency into a wholly positive force—a healthy force that did more than just reduce negative effects.

Shipping material represents about the most utilitarian good imaginable. One might have trouble getting animated about the destructive force of mail, but consider the truckloads of materials each year (660 million pieces of mail moved a day), vast amounts of toxic ink, laminate coatings, and glues that are exposed to humans through direct contact and later dumped in a landfill. If the USPS could make safer postal products, it could make a safer world.

As we said earlier, there are two ways to create a safe, healthy product: either deconstruct an existing one and replace the dangerous materials in it with good ones, or start with a known list of positive materials and construct the product. For an entity like the USPS, taking apart each envelope, box, piece of shipping tape, label, and so forth proved to be the more viable course.

How complicated could that be?

Well, as it turned out, pretty complicated.

For the five product categories MBDC evaluated—paperboard envelopes, corrugated fiberboard (aka boxes), standard plastic envelopes, pressure-sensitive mailing labels, and pressure-sensitive tape—there were 175 suppliers and 200 materials composed of some 1,300 different ingredients.

The USPS committed to understanding every single material in its supply chain and to phasing out, right away, the highest-priority substances—namely, anything teratogenic, mutagenic, or carcinogenic (our "red" list, under certification protocol). Advancing through a progression of inventory, assessment, and optimization, we looked at individual chemicals and finished products from environmental and human health perspectives. We broke down each substance in the supply chain.

While it's all information-based, all science, it's not as if we put the product in a chemical-analysis box, pressed a button, and out came everything we needed to know. The process was deliberate and painstaking—and still is—and involved looking at the substance or product down to the root. After we did that, we considered performance, cost, and aesthetics.

We analyzed a cardboard mailer that used two or three different types of adhesive. The cardboard also included a starch, a colorant, and several other materials. While assessing the mailer's release liner—the strip you pull to seal a mailer—we found that the ink (used in the instructions for how to remove the strip) contained an organofluoro compound (PTFE)—a problematic substance. The PTFE was there for scratch resistance, to prevent the ink words from coming off. Through our optimization work, we found a replacement substance that is acceptable from our chemical-standards point of view and gives the same performance.

Our work with USPS eventually expanded to stamps and stamp products. Even before meeting with our design team, the USPS had done a lot of work to equip stamps with a benign adhesive, so the stamp would not interfere with recycling the base material (the postcard or envelope to which it was affixed). But stamps have extremely high performance requirements: special inks that show up under UV light to protect against forgery, for example. Working together, we made those clean too.

Along with this close work is the considerable job of educating suppliers—in the case of the USPS, a behemoth chain. Sometimes it's a challenge to convince the suppliers to provide data. Their formula is a proprietary mixture, after all, their differentiator, the way they make money. Replacing a supplier who doesn't wish to assist this Cradle to Cradle process with another supplier is not so simple; in many cases, the client has a solid, long-standing relationship and has developed a friendship of sorts. This can, in fact, be the basis for getting a supplier to consider making a change.

On the other hand, less crucial suppliers might be changing out quickly, all the time. One of MBDC's functions is to reeducate and reconfigure (glues or inks or production) with new fabricators on an ongoing basis. We originally worked with 20 suppliers; more than 250 have now been part of the process.

Although this larger network of suppliers obviously makes the work more labor-intensive, what a tremendous opportunity! Each new and old vendor now knows that the system can be different and that it can take that knowledge to other customers with a potential cascade of future positive effects.

We want to point out a key fact: This work we did with the USPS was originally revenue-neutral for them—meaning it did not cost them more to create Cradle to Cradle Certified^{CM} postal products.

If you stop to think about it, that might seem a bit astounding. All of the products we helped create with them have been fully defined and redesigned to work within the biological and technical food chains, and they are no more expensive than the original products they replaced. Cradle to Cradle is essentially competing with highly efficient, uncompostable past design—it might have taken decades to get the uncompostable ink to be as inexpensive and effective as possible—and matching that efficiency with compostable inks and papers and glues. Innovation didn't cost extra.

As these materials and strategies are deployed, as the system grows, it becomes even better. It can be revenue-positive. The companies that embark on this work inspire whole other industries to take up Cradle to Cradle materials and techniques; the cost for these replacement components or ingredients drops because of increasing scale of production and instant knowledge of what healthful materials can be utilized. An ink manufacturer, for example, now knows how to create safe ink and can market it to other producers. Or another ink manufacturer adopts the formulas—what we are calling technomimicry. When we advised Goodbaby, we were able to use the Cradle to Cradle method of foaming aluminum that we had helped discover for Steelcase, wherein a process native to the plastics industry was applied to metals. Such technomimiory has far-reaching effects when one considers the quantity of products created by Goodbaby every year. Which industry can use that next?

Cradle to Cradle also essentially simplifies design. Take the television example again. Michael replaced all off-gassing plastics in the television, reduced the overall number of chemicals, and eliminated PVC. Simply reducing the number of chemicals employed lowers the cost of production—of finding the suppliers for such a variety of substances, of purchasing them, of storing them, of hiring employees who can handle so many different materials. Philips also rethought packaging from a Cradle to Cradle perspective, and the company saved €20 million a year.

But to return to the USPS, the agency continues to work with us to improve its formulations. It's not always easy to be an industry leader in this way. Even simply figuring out how to send the value statement through the whole chain of command can be complex, given the scale. But the legacy of this work endures in all of the design solutions we have discovered together, for the USPS in the future and for industries around the world.

Likewise, when Bill and Brad Pitt cofounded Make It Right, to assist in the rebuilding of the Lower Ninth Ward in New Orleans after Katrina, they set the intention to create a neighborhood of Cradle to Cradle–inspired buildings, with the holistic consideration of energy, materials, water, and site. We asked donors to contribute funds to help bring displaced families home. Why would anyone want to return to an unhealthy home? With Make It Right, we decided to engage two dozen diverse local and international architects. We provided guidance on the Cradle to Cradle approach. Of course, not every material and technology solution employed in the Lower Ninth is perfect, but we are working toward that goal, inspired by a whole host of manufacturers—and the architects themselves. The result, as of the fall of 2012, is more than 80 homes built with LEED Platinum ratings. The average monthly electric bill for residents plummeted from hundreds of dollars to as low as $40. The children living in these homes are healthier. One girl, who had loved to dance but was prevented by asthma, returned from Texas to her family's new Make It Right house. Her condition improved to the point where she could go to the dance classes her mother could now afford and pursue her dream.

That seems to us a wondrous butterfly effect of stating the goal of making Cradle to Cradle–inspired design decisions and using healthful materials whenever possible.

The butterfly effect ripples to whole towns and countries. The Netherlands has been embracing Cradle to Cradle concepts with vigor. Entire regions have adopted the principles. The province of Limburg, for example, is applying Cradle to Cradle to renewing urban areas and is collaborating with companies on creating better products. Venlo in particular has embraced the Cradle to Cradle concept with the development of Greenport Venlo, a business district that hosted a world horticulture expo, the Floriade 2012.

Cradle to Cradle for Everyone

As of 2012, there were 70 million known organic and inorganic substances in the world. We alone, working with our clients, even with their magnitude, can't get it done. It's going to take us all, and it's going to take us forever. We need that multiplier effect.

When it comes to moving toward a Cradle to Cradle world, we not only welcome the butterfly effect, we actively *need* the butterfly effect. We need this knowledge and process to ripple out across industry. That's why we were so excited to launch, in 2010, the Cradle to Cradle Products Innovation Institute in San Francisco (c2ccertified.org).

As the institute develops, it can help companies, organizations, and governments create meaningful goals, because road maps will finally be available. For example, a city, state, or country could set a goal of certifying all its paper through the Cradle to Cradle protocol by 2020. Why and how would they do that? Because our processes would help identify and connect all the manufacturers and suppliers who are optimizing, and that information will be accessible through the institute.

We have met with paper and ink manufacturers who can't or won't spend the money for a new and unfamiliar certification, and we understand this from a business point of view. But what if we created a set of product-innovation protocols as templates for continuous improvement—one for publishers of newspapers, one for magazines, one for books? When you use this protocol, you know you are satisfying the triple top line of growing economic, ecological, and social benefit. You know it is easier to achieve the goal. And it is an evolving package; when, for example, a new, better optical whitener becomes available, it could replace the old one. All of this information will be available for all to use.

It's worth noting that Van Houtum B.V. manufactures the first, and so far the only, toilet paper in the world that can be safely disposed of—it leaves a beneficial poop print. The soaps and toilet paper are made without using harmful chemicals (replacing 29 of 30 chemicals used in traditional toilet papers), so they're safe to use and safe to dispose of. The same applies to their hand towels, which they collect and reuse from businesses, airports, and even the Floriade 2012. By putting up simple signs that ask that people not throw gum, lipstick, or other paraphernalia into the waste bin, Van Houtum recovers nearly all of its paper, ready to make it new again. It's an excellent example of a company that has high revenue streams, is incredibly successful, and at the same time has created a beneficial product.

About the Cradle to Cradle Products Innovation Institute

The Cradle to Cradle Products Innovation Institute, a 501(c)(3) nonprofit, is an international organization that the two of us founded to bring about a large-scale transformation in the way we make things.

Our goal for the institute is to scale up the application of the Cradle to Cradle Certified^{CM} program, which recognizes transition and human intention as part of any successful protocol for continuous product improvement. The protocol guides product development across five interrelated quality categories:

- Making products out of materials that are safe and healthy for humans and the environment;
- Designing products so all materials can be reused by nature or industry, including reverse logistics systems;
- Assembling and manufacturing products with the use of renewable, nonpolluting energy;
- Making products in ways that protect and enrich water supplies;
- Treating all the people involved in a socially responsible way.

We licensed this standard—the Cradle to Cradle Certified^{CM} program—as a contribution to the institute after 15 years of private development with many of the world's top brands. Ongoing development and improvement of the standard are governed by an independent Certification Standards Board, which is composed of leaders from academia, the NGO environmental community, government, and industry.

The institute's mission is to provide this continuous-improvement quality standard to guide product manufacturers and designers in making safe and healthy products that are continuously reusable in biological and technical nutrient cycles, and have been created using clean energy, clean water, and social fairness.

Continuous improvement with this systemic approach spurs the creation of truly innovative, high-quality, and beautiful products and transforms their production into a positive force for the economy, society, and the environment. The certification mark provides consumers, regulators, employees, and industry peers with a clear, tangible understanding of a product's sustainability.

The institute trains and accredits assessors who assist companies in preparing for product certification. The assessors submit their results to the institute for auditing, and a certificate is issued if the product meets the prerequisites.

The institute is governed by an independent board of directors and is led by Bridgett Luther, former director of the California Department of Conservation. It is headquartered in San Francisco, California, with a satellite office in the Netherlands. Learn more at c2ccertified.org.

Speaking of paper, that brings us to this very book, the one you are holding in your hands (the tangible version of the book, that is). In 2002, limited paper options inspired us to conceive *Cradle to Cradle* as a technical nutrient. Making a book from new materials—polymers—was easier to do than to optimize the paper! People have been using paper for centuries, but printed materials have never been designed to fit into biological cycles. For nearly two decades, we've been working with paper, ink, and glue manufacturers to optimize the process, to make books into biological nutrients.

We had hoped to be able to suggest that, after reading *The Upcycle*, you could safely eat, plant, or compost it—or, if you were so inclined, burn it. The teams at MBDC, EPEA, and Melcher Media spent years researching paper, inks, glues, and manufacturing techniques to come up with the book you hold. The result represents a place on the continuum of continuous improvement—it's not the perfect solution. However, it represents the upcycling of the printing industry. For now, we hope you will keep it or pass it to a friend; but if you want to discard it, a traditional recycling process will work.

First, we found a special printer, Quad/Graphics, whose team tirelessly collaborated with us to examine new materials and allowed us to specify almost every step in the process. For paper, the most optimized solution we discovered in the United States is a sustainably sourced, virgin paper, manufactured by Glatfelter using about 50 percent biomass for energy. Made with no added optical brighteners or elemental chlorine bleaching, it minimizes the presence of chlorine, heavy metals, and other problematic chemicals found in typical recycled paper, and it will contribute to a cleaner recycled paper stream. It is safe for combustion without a filter and almost completely compostable except for one chelating agent included at the infinitesimally low level of ten parts per million. This chemical is nontoxic and does not accumulate.

However, it does not degrade in compost. We are working on optimizing it now—the next edition will be better.

For the cover, we chose C-Stone's Bio-Paper, a blend of plant-based resin and limestone made in Southern California. For the binding, we selected an adhesive with the safest formulation we could find. The cleanest ink we identified is produced in Europe by Flint Group, who is working with us to design a biological nutrient ink.

We felt it was important to both source the paper and print this book in the United States, to avoid shipping pallets of books overseas; better for the optimized ink (which doesn't weigh very much) to travel from Europe. Soon, we hope there will be sources for these biological nutrients everywhere. It's all about continuous improvement.

Strategy of Support: Learning to Run Together

We've been moving as fast as we can across rough territory. We don't want you to think that moving your company, your home, your community to upcycle is going to be easy. Sometimes we fall down and get back up. But we have learned a key component in overcoming the inertia. We all need help from others along the way; finding mutual support is important. You need to figure out whom you are dealing with when you go into a situation, and you also have to know how people tend to react to change.

So let's talk about Icelandic horses.

As background, we'd like you to consider a fascinating study from the early 1980s by two dozen psychologists who tracked about 2,500 people over 25 years for changes in personality and the causes. It was interesting to note that the most frequent cause of personality change was financial stress. Emotional stress was a distant second. The study didn't actually check what *didn't* change under any circumstance, but when we called the researchers, asking them to review the data for such information, they pointed out that there were two immutable personality characteristics, no matter the circumstances, that didn't change: friendliness and willingness to experiment with new things.

Now we take you to Iceland. The country has gorgeous small horses, known for their loyalty, unique gait, and ability to run on rough terrain. In Iceland, people don't break the horses. The horse trainers and riders spend time with the horses and watch them, looking for particular characteristics. When asked what it is they are looking for, the answer is friendliness and willingness. They want the horses that are happy when a person is riding them; they want horses that are friendly enough to maintain themselves as their own characters while allowing a person to climb on their backs, and they must be "willing" and love to run, which makes the relationship fun. They especially like the *toelt*, a smooth and comfortable running gait.

In Iceland, there is an intense natural selection process involved here: If the horses are not friendly and willing, Icelanders eat them.

So that brings us to a large meeting we were leading at a major U.S. manufacturer. At one point during the meeting, we realized that there were seven different types of people in the room. There were seven ways people were organizing themselves around change. Essentially, we had recognized a scale of human friendliness and willingness.

This helped us going forward. You too might find this insight helpful, so you can be prepared and not get discouraged when you try to spark innovation in your workplace or community. Here are these personalities, from the enthusiasts to the curmudgeons.

Ones are the 100 percent acolytes who hear Cradle to Cradle and upcycle ideas—go renewable, or go safe and healthy—and are delighted to engage and want to devote their lives to it. These are the extremely willing, the extremely friendly: *This was fabulous. My life has changed forever.*

The Twos think these ideas will make their jobs more exciting, and since they are dedicated to their jobs, this is an improvement in their lives. These people don't walk out of the meeting all dreamy and head for Yosemite to think about it. They like how these ideas give their job a purpose, a frame: *I will be so dedicated to this idea.*

The Threes are a bit more sanguine. They think: *This is really interesting, and now I will get to do my job this new way. It makes my job better inasmuch as I can go home and tell my kids, "Today I solar-powered a factory."*

The Fours are willing to do their jobs in a different way, as long as the boss lets them go home on time and they've got a retirement plan: *Fine, if this is what is expected of me, I will gladly do it. It's just a job.*

The Fives couldn't care less one way or another. They just want to do their jobs and go home. They have no emotional commitment to the new framework nor any intellectual curiosity: *Whatever.*

Then you have the Sixes. They find all of this talk of change—of production or manufacturing or systems—disconcerting. They are often in mid-level management, sometimes very metrics-and-data-focused, and are troubled that, if indeed this is the way of the future, they may have been doing wrong things for years. They have to justify their careers and their positions as seniors in their organizations. They may become what we call "deep-sixers." They try to deep-six, to sabotage, the project by passive-aggressive technique—somehow the memo doesn't get out on time, or the information is not shared properly. One of the biggest logistical difficulties in moving to a Cradle to

Cradle world is the immovable people. We sometimes find ourselves wondering if these people think of themselves as mules. The thing to remember about mules is that while they are very strong and useful, if sometimes also hard to get moving, they are also a hybrid of a donkey and a horse. But we remind ourselves: They are one generation. They are sterile. We wave goodbye. It's not worth fighting. We sometimes have to ask them to step aside from the team for a time (and join in when they are ready) to maintain momentum in a project.

The Sevens, on the other hand, are the active aggressors, direct antagonists. We truly appreciate and require them. They test us and everyone else. We end up in the Nietzschean situation of "what does not kill us makes us stronger." They are wonderful; they are honest. And once they are personally convinced and get on board, they are very powerful advocates for an upcycled world. They become Ones.

Start Where You Are: No Fear

And now we come to you. Our first book seemed to inspire a good deal of positive feedback, but we also heard from readers who felt overwhelmed by the magnitude of the shift we were proposing. In some cases, the message they heard was "Change everything you're doing! Right away!"

An interior designer told us that while attending a Cradle to Cradle workshop, she kept saying to herself, "How can I continue being a designer knowing that all of this beauty is made of toxic materials? How can I, in good faith, order new carpets, drapes, and furniture now that I understand the negative impact they are having on our air, land, and water? Am I going to have to quit being a designer and become an activist?!"

We don't want people to feel overwhelmed; the upcycle is about a shift to a delightful, safe, healthy abundance. There's nothing delightful about panic and fear. Individuals, companies, and communities that aren't able, or ready, to take huge steps or redesign a product or system completely right now can begin the upcycle in a variety of smaller ways to get started in the right direction. Upcycling is moving positively forward, no matter how small the steps. So what are our suggestions for this designer and others like her?

The first step is to make it clear to manufacturers and suppliers up the chain that she prefers defined products. If she finds a beautiful fabric but doesn't know what chemicals constitute the colors, there are ways to begin finding out.

One way is to go to the Cradle to Cradle Products Innovation Institute website to find out which products have been assessed, which banned substances have been identified, what materials are suboptimal, and what materials are defined and safe in the environment in use and reuse cycles. She can consult the institute's data sets and find the specific products that have been certified or have begun the process. If she can't find what she needs, but she wants a company to improve its fabric or another product, she can send a signal to it through the online wish list we described before. She can say that she needs Cradle to Cradle–optimized drawer pulls, for example. Or a Cradle to Cradle–certified shade fabric.

Eventually, she can download templates on how to query industries about what chemicals are in their products. She may then ask the supplier of a gorgeous fabric from France, "Where do these reds come from? Is there anything to be worried about?"

If the manufacturer and supplier aren't able to provide the information she needs, she can ask that they find out and encourage others with similar concerns to make a request as well. Assuming it's interested in improving its product and its relationship with customers, the fabric company can turn to assessors who are trained in our methodology and understand the subtlety of the inventory, assessment, and optimization protocols. These assessors can then submit their evaluations to the Cradle to Cradle Products Innovation Institute in order to earn the use of the certification mark if the customer wishes.

Does this mean that our designer friend has been able to instantly perfect all of her materials and her processes? Probably not—but she's started in the right direction by recognizing and signaling her intention of what she would like to do. Obviously, for the upcycle to work, to improve systems and products that already exist, we have to be able to start where the designers are, in real life. And we all can do this piece by piece, day by day.

When asked if the historic *Apollo 11* mission to the moon was ever off course, astronaut Buzz Aldrin replied, "It was all course correction." They could see the moon and continuously improved their trajectories. We ourselves have been correcting course and have learned an enormous amount about what works with the Cradle to Cradle strategy and what does not. We have sat around tables with early adopters, from different industries, and listened to the challenges they face in implementation. We have used the valuable feedback of people out in the field. We want to give back that knowledge as idea nutrients, to get everyone thinking about a world of upcycle. And we will be continually updating that information for you.

Love All the Children, All the Time

The great evolutionary biologist E. O. Wilson calls humans "the future-seeking species" and suggests that natural selection has made hopefulness a unique human quality, "a necessary companion of intelligence." The philosopher Gottfried Leibniz said, "Everything that is possible demands to exist." The economist Kenneth Boulding said, "What exists is possible." Though circumstances may seem dire at times, what we have already achieved, what we have already seen to be possible, can inspire our vision for the future. That vision cannot be merely less tragic. There is a credible vision out there in which our species does more than just survive. It thrives.

We could chastise people, including ourselves, for the damage wrought on nature in the past, or for our species' suboptimal designs—but what's done is done. Would it not be more energizing to start celebrating a system that is coherent, powered by the sun, truly cyclical, and therefore abundant?

We invite you to join the growing upcycle community, both through established networks and online, to bring the ideas within it, and your own ideas, to life. If we can engage the designer that lurks in each and every reader—business and governmental leaders, architects, planners, engineers, entrepreneurs, policy makers, product designers, students, educators, parents, and citizens of the world—what some may see as the utopian vision we have painted is within our grasp.

The seeds of transformation may be found within these pages and within you. As the natives in the Amazonian forest did when they were moving through, let's start to choose those things that as a species we wish to see on the planet, for all of us here now and the many more who will be here in 2050, leaving a positive footprint.

Gouverneur Morris, author of the preamble to the U.S. Constitution, used the beautiful and heartbreaking phrase "a more perfect union." This nimbly captures the nature both of what one strives for and the eternal, and motivating, elusiveness of the striving. *"More perfect"* is the upcycle.

The goal of this book, then, is twofold: to express the goal, of course, but also to invite and inspire you to take these ideas and make design revolutions of your own. There are countless initiatives going on right now worth noting, publicizing, joining, emulating. We know we're not the only ones.

As to the ideas we've outlined here: Take them—please. Turn them into realities. They need your creative interpretation, your local application, your passion. Log on to the Cradle to Cradle Products Innovation Institute website and find like-minded people with the resources you need, your community of support. We work with hundreds of universities, private-equity companies, venture capitalists, and manufacturers. Seek out the ones that can help you and your opportunities will appear. "Seek constant improvement by the sharing of knowledge," states the ninth and last of our Hannover Principles. "Encourage direct and open communication between colleagues, patrons, manufacturers, and users."

It's famously said, "Think global, act local." But within the Cradle to Cradle framework, we think it might be possible to think universal and act molecular. Your considerations can be expanded simply by the information available today, to take account of the needs of the entire natural world—pointedly, including us humans, including every element.

We will leave you with a few last words, but we hope to hear from you again, either through the Cradle to Cradle Products Innovation Institute, via our website, or when we read an article online about how you solved the challenge of resequestering carbon from the atmosphere, or devised a nutrient management program for your block, or figured out how to design a fabric that uses optics instead of dyes formulated with heavy metals to achieve vivid colors.

Obviously this is big work. Getting to a Cradle to Cradle world will require most everyone's contribution. Everyone has different qualifications and different talents, thankfully. We will need that diversity. No one wants to be a zero. Fortunately, no one is.

What's
Next?

Chapter 7

If you take anything from this book, we'd like you to remember the 10 points that follow. We think they may be beneficial for your work and your life, as they have been for ours.

1. We Don't Have an Energy Problem. We Have a Materials-in-the-Wrong-Place Problem

Carbon is perfect. It is crucial to human life. We need it on earth.

Unfortunately, it is now in our air and water in overabundance, where we cannot utilize its strengths. We need to reconfigure our systems to keep carbon earthbound.

Fortunately, there are many ways we can do this. We can use fossil fuels for key goods, such as medicines, while we use renewable energy for power. We can sequester carbon emissions from biodegrading materials and use them to create biogas and soil nutrients. The carbon goes back into the earth where it belongs.

Once we reorganize, we will grow—literally.

Likewise, we don't have a toxins problem; we have a materials-in-the-wrong-place problem.

If we realize that what we essentially have is a sorting problem, we can begin the process of reorganizing so we never have to worry about these issues again.

At the beginning of the book, we said to toss out the idea of a nurturing "Mother Nature." No need to romanticize nature. Nature is not exclusively benevolent. But nature *is* fairly intelligent after millions of years of evolution. For example, nature evolved to put in mother's milk exactly what is needed to nurture and grow new life; it works.

Recently, we have added many new ingredients to breast milk—bromide-based fire retardants, for example. We can look to nature not as our mother but as our teacher. Nature gave us the correct recipe. If nature didn't put bromide-based fire retardants in milk in the first place, we have no reason to add it. It is a materials-in-the-wrong-place problem. Let's redesign.

2. Get "Out of Sight" Out of Mind

You don't have a garbage can. You have a nutrient rest stop.

You don't have a toilet. You have a nutrient cycling system.

Think and act on the idea that when you flush a toilet you are sending beneficial nutrients to a processing plant to be made into the key ingredients needed by farmers to grow healthy food for you and your family.

Get greedy about your garbage. Now that the world has started down the path of upcycling, plenty of companies covet what you put in the trash can every day. You can value it too. Instead of asking yourself, "How do I get rid of this?" ask, "How much money could I get for this? Who could enjoy the benefits of these great nutrients? My city, my neighborhood, my favorite nonprofit?"

Next time you want to use the word "waste," bite your tongue. Worms consume food and, through the system of their bodies, produce richer nutrients. You, through the system of your intelligence, can create richer nutrients too.

3. Always Be Asking What's Next

In *Cradle to Cradle*, we confused some readers with the term "closed loops."[14] We intended to mean that a material or its component chemicals could be reused endlessly, safely. But the term allowed the misinterpretation that it was okay to design a toxic product in the first place, as long as it could be reconfigured into another toxic product. We never meant that. That leads to monstrous hybrids.

We want you to always think, *What's next?*

See "Closing Some Loops" on page 43.

What will happen next to the shirt I design today? What is next for this book?

We want you to think of every component of your design as being borrowed. It will be returned one day to the biosphere or technosphere. It is your role to return it in as good a condition as you found it, as a good neighbor would. You have that chemical or heavy metal in usufruct. You have that chemical or heavy metal for a reuse period, and then it moves on to another product without tainting the biosphere or technosphere. Design for your particular reuse period, always with its next reuse and its next reuse and its next reuse in mind.

4. You Are Alive. Your Toaster Is Not

We have been in this work for decades and still . . . *still* we stop every time a company mentions a technical product as having a product life or life cycle. This means how long a product is used by the customer. But the term confers a kind of superiority on the product that it doesn't deserve. The technical product is not "alive"; it is inert. It also suggests that the product will die and go away and never bother us again.

But technical products don't die and vanish. This is the problem and the opportunity. Products stay on and on and on. Maybe as toxins in a dump. Or a plastic bottle cap bobbing in the ocean. We need to get away from thinking of these objects as mutable or we won't consider their endless reuse. They are technical nutrients. We can use them over and over. We can design them to be part of a Materials Bank and lease the steel and rubber in a washing machine, for example, until the time comes to use them for something else.

Conversely, there are such things as people, and they *are* alive. Companies talk about human resources departments, as if people were just commodities owned by the company, goods to be used. No. We are people. A better term might be human relations departments, since they can focus on how the company is relating to the needs and desires of the people who work so hard for them, so as to create an optimized relationship.

5. Optimize, Optimize, Optimize

Speak to the world in positives. "We will run on 20 percent renewable power by 2020 and 100 percent as soon as it is cost-effective." "We will use 95 percent of our gray water in 2015 and 100 percent by 2017." "We have made a product that, when used, provides part of a person's daily requirement of minerals."

It doesn't make us happy to see your downward-sloping lines of fewer carbon emissions, fewer toxins. We want to see your rising lines of positive aspirations and beneficial commitments.

You are doing good. Enjoy it. Say it. We like to hear it. Upcycle your descriptions of your work and progress. Don't be a pessimist. *The glass is half empty.* But don't just be a passive optimist either. *The glass is half full.* Start with inventory; take scientific stock of your situation. *The glass is full of water and air.* Then signal your intention for design. *I want the glass to be bigger.*

6. You *Can* and You *Will*

No need for scolding. No need for "shoulds" and "musts." From our work around the world, we have come to see that human beings are essentially in agreement on what is needed to integrate ourselves into the natural upcycle of life. Businesses and governments understand these needs too. What gets in the way is fear, fear that the changes and innovations are impossible, or too costly, or we don't have enough information—there are a million reasons not to change.

The job of upcycle advocates is to encourage people and to inspire behaviors, helping all entities understand that change is possible, beneficial, profitable. The city *can* look into creating biogas plants at the local dump to create free energy. The company *will* resell its used paper to a nutrient manager who uses the materials for other key products and pays the company back.

This is a joyful project before us. Let's speak that way too.

7. Add Good on Top of Subtracting Bad

In the organic food world, a farmer can begin growing organic at any time, but he or she cannot get organic certification for three years. Why? Because you can't just throw pesticides on your fields one year and then stop using pesticides and think they have worked their way out of the soil. Meanwhile, in the three-year wait time, the dedicated farmer might be spending more money setting up these new farming techniques, yet he or she can't get the financial benefit of the higher values associated with organics until three years out.

We can find ways to honor people's intentions.

When we certify in our Cradle to Cradle Certified[CM] program, we begin by certifying intentionality. No one can get to perfection overnight. But people can be honored, recognized, and encouraged for having begun in earnest.

Starting is important. And creating additionality is essential. If we want sustainably harvested wood, we not only need to source from sustainable forests, we need to create new sustainable forestry programs too. But you can start where you are today—if you have planted even one tree, you deserve to be recognized.

Renewable energy benefits from additionality too. We know the sentiment is good if a company buys offset credits for its carbon emissions, but upcycling is happening if the company is participating in the creation of new, local renewable power sources that wouldn't have been there without its instigation. Think small, think big, think adding good on top of subtracting bad. There is always room for more additionality. We can add on, not just pile on.

8. Gaze at the World Right Around You . . . Then Begin

In *Cradle to Cradle*, we talked about the need to build for the local community. You don't want to be in Bali in a sealed-up, air-conditioned building designed for Shanghai, Chicago, or Frankfurt, for example. It wouldn't be appropriate or delightful in the context of Bali. But we can think about locality in many ways. When helping create Sustainability Base for NASA, we thought of the locality— of its resources (good sunlight, breezes for cooling and refreshing air)—to design the building. When proposing a green roof for the Ford Motor Company plant, that was an answer to a specific local need of containing storm-water runoff. On a building in Arizona, it might be better to create a solar roof or a rooftop greenhouse recycling precious water and nutrients.

In Barcelona, we didn't propose to use nonnative rain forest fig trees in the pharmaceutical laboratory atrium. We designed a re-generator for the butterflies that were local to that city, endangered butterflies that needed support and celebration.

Get specific about your locality. You will arrive at more ingeniously indigenous solutions if you let the locality guide you. Some solutions can have global benefits and applications, but remember to start where you are. All sustainability, like politics, is local.

9. The Time Is Now

We—all of us—have a lot to do. We know that this work requires all of us and it will take forever, but some of this work is urgent. Rebuilding the phosphate in our soil is crucial and, as we have pointed out, the solution easily within our reach.

Rebuilding soil in general is as important right now to our future as converting to renewable energy; soil is currently disappearing faster than we can restore it. No time to waste, watching it blow away. The sooner we start recharging our soil battery, the sooner we can grow.

We need to focus on indoor air quality. In much of the world, people spend 80 percent of their time indoors and yet the air is often worse at their desks than outside on the streets. We can focus on every material introduced into our homes and office buildings to make sure we are breathing only what our bodies enjoy.

And we can immediately stop introducing unknown chemicals and materials into our biosphere. We can't afford this Russian roulette. We can't wait to learn whether or not these chemicals will harm us in the long term. Let's stop using them in the short term.

Let's start now. The precautionary principle is and is about being alive and well.

10. Go Forward Beneficially

You have one life and, like a tree, you can create abundance, a profusion. You can celebrate your emissions. Every year of your life, you are accumulating more potential for good for the world. We know that with your intelligence, your talents, your intent, you will make life for your contemporaries and for future generations better.

You are a known positive. No need to think of yourself as misplaced in the natural world, or that you cause destruction with your presence. You can contribute. You are part of the ever-upcycling path of life.

Accept that deep in your heart and mind.

Then go forward. Be successful.

We hope to enjoy all that you share.

And tell your children that things are looking up.

Notes

Notes

Chapter One. Life Upcycles

27 **"the word 'wilderness'"**: René Jules Dubos, *The Wooing of Earth* (London: Athlone Press, 1980), 10.

32 **The entomologist E. O. Wilson:** E. O. Wilson and Bert Hölldobler, *The Superorganism: The Beauty, Elegance, and Strangeness of Insect Societies* (New York: W. W. Norton, 2009), 5.

33 **"The vegetable life":** Ralph Waldo Emerson, *The Works of Ralph Waldo Emerson, Volume 1* (New York: Hearst's International Library, 1914), 358.

47 **what Claude Lévi-Strauss wrote:** Claude Lévi-Strauss, *A World on the Wane*, trans. John Russell (New York: Criterion Books, 1961), 127.

Chapter Two. Houston, We Have a Solution

58 **The plane flew:** Mike Hirst, "America's Man-Powered Prizewinner," *Flight International*, October 29, 1977, 1254.

Chapter Three. Wind Equals Food

93 **Now the poachers help:** Prakratik Society, *2004 Technical Report, the Ashden Awards for Sustainable Energy* (India: 2004).

94 **The Peruvian economist Hernando de Soto:** Hernando de Soto, *The Mystery of Capital: Why Capitalism Triumphs in the West and Fails Everywhere Else* (New York: Basic Books, 2000).

95 **When humans use petroleum:** Thom Hartmann, *The Last Hours of Ancient Sunlight: The Fate of the World and What We Can Do Before It's Too Late* (New York: Three Rivers Press, 1998), 7.

97 **In the United States:** United States Nuclear Regulatory Commission, *Congressional Budget Justification: Fiscal Year 2011* (Washington, D.C., 2010).

97 **cost of building nuclear reactors:** Pilita Clark, "Nuclear 'Hard to Justify,' Says GE Chief," *Financial Times*, July 30, 2012.

98 **plausible ways:** Mark Z. Jacobson and Mark A. Delucchi, "A Path to Sustainable Energy by 2030," *Scientific American* 301, no. 5 (November 2009): 58–65, accessed November 5, 2012.

98 **"We are like tenant farmers":** James Newton, *Uncommon Friends: Life with Thomas Edison, Henry Ford, Harvey Firestone, Alexis Carrel & Charles Lindbergh* (New York: Mariner Books, Houghton Mifflin Harcourt, 1989), 31.

99 **powers 66 percent:** Orkustofnun, National Energy Authority of Iceland, "Geothermal," accessed November 6, 2012, www.nea.is/geothermal.

100 **Europe's importation:** "Commodity, Crude Palm Oil," CRN India, www.crnindia.com/commodity/palmoil.html.

101 **an effort by the Icelandic government:** Mark Lynas, "Damned Nation," *The Ecologist*, January 1, 2004.

101   **100 percent of the country's electricity:** Jessica Aldred, "Iceland's Energy Answer Comes Naturally," *The Guardian*, April 22, 2008, www.guardian.co.uk /environment/2008/apr/22/renewableenergy.alternativeenergy.

104   **cost per kilowatt-hour of wind:** Mark Bolinger and Ryan Wiser, *2011 Wind Technologies Market Report* (Oak Ridge, TN: U.S. Department of Energy, 2012).

106   **Juhl's solution:** Interview with Dan Juhl, August 2012.

108   **another energy solution:** "Biomass Energy: Manure for Fuel," State Energy Conservation Office (Texas), last modified April 23, 2009, www.seco.cpa.state .tx.us/re_biomass-manure.htm.

108   **Axpo Kompogas . . . fermentation processes:** "Axpo Kompogas Ltd. Company Profile, Energy From Organic Waste," Axpo Kompogas, Ltd., www.axpo-kompogas .ch/files/artikel/99/Axpo_%20Kompogas_%20Ltd_Company_Profile.pdf.

113   **spending hundreds of millions:** "Secure Border Initiative Fence Construction Costs," U.S. Government Accountability Office, accessed January 11, 2012, www.gao.gov/products/GAO-09-244R.

114   **electricity must travel:** "Solutions for Electricity Loss," EcoWatt Energy, http://ecowattenergy.com/blog/2011/solutions-for-electricity-loss/.

115   **Walmart is committed:** "Global Responsibility, Environmental Sustainability, Energy," Wal-Mart Stores, Inc., http://corporate.walmart.com/global -responsibility/environment-sustainability/energy.

117   **the price will come down to compete with CFLs:** Aaron Ziv, "Lumens vs. Watts for LED Bulbs," *National Geographic, Green Living*, accessed November 6, 2012, http://greenliving.nationalgeographic.com/lumens-vs -watts-led-bulbs-2789.html.

117   **the surface area of Lake Mead:** "Hydropower at Hoover Dam," United States Department of the Interior, Bureau of Reclamation, last modified February 2009, www.usbr.gov/lc/hooverdam/faqs/powerfaq.html.

117   **the estimated energy needed:** "How Much Electricity Is Used for Lighting in the United States?" United States Energy Information Administration, last modified January 9, 2013, www.eia.gov/tools/faqs/faq.cfm?id=99&t=3.

Chapter Four. Soil Not Oil

126   **depleted 75 percent of its topsoil:** "Land Degradation," Global Change Program, University of Michigan, last modified January 4, 2010, www.globalchange.umich.edu/globalchange2/current/lectures/land_deg /land_deg.html.

126   **the Iowa prairie:** Patricia Muir, "Erosion," Oregon State University, last modified October 5, 2012, http://people.oregonstate.edu/~muirp/erosion.htm.

127   **alternate green revolution:** Sir Albert Howard, *An Agricultural Testament* (Oxford, U.K.: University Press, 1940).

128   **Gary Zimmer:** Gary Zimmer, *The Biological Farmer: A Complete Guide to the Sustainable & Profitable Biological System of Farming* (Austin, TX: Acres U.S.A., 2000).

130   **the United States' supply:** C. Robert Taylor, "Forget Oil, Worry About Phosphorous," *Daily Yonder*, last modified September 13, 2010, www .dailyyonder.com/forget-oil-worry-about-phosphorous/2010/09/08/2929.

130   **the recycled water:** Felicity Barringer, "As 'Yuck Factor' Subsides, Treated Wastewater Flows From Taps," *New York Times*, February 9, 2012.

131   **NEWater now accounts for:** "NEWater," PUB, Singapore's national water agency, last modified July 9, 2012, www.pub.gov.sg/water/newater /Pages/default.aspx.

135   **feeding chickens arsenic:** Darryl Fears, "Maryland Set to Become First State to Ban Arsenic in Chicken Feed," *Washington Post*, April 9, 2012.

135   **permaculture experiments:** "Update on the Jordan Valley Permaculture Project (aka 'Greening the Desert: The Sequel'): Leave All Expectations Behind," The Permaculture Research Institute of Australia, http://permaculturenews .org/2011/02/19/update-on-the-jordan-valley-permaculture-project-aka -greening-the-desert-the-sequel-leave-all-expectations-behind/.

137   **agricultural exports:** "Agriculture and Food," NL, a division of the Ministry of Economic Affairs, www.hollandtrade.com/sector-information/agriculture -and-food/.

138   **saved the building $5,000 per year:** "Chicago City Hall," The Green Roof Projects Database, www.greenroofs.com/projects/pview.php?id=21.

138   **A former naval yard building:** Lisa W. Foderaro, "Rooftop Greenhouse Will Boost City Farming," *New York Times*, April 5, 2012.

139   **increased vegetable production:** Amy Bentley, *Eating for Victory: Food Rationing and the Politics of Domesticity* (Urbana: University of Illinois Press, 1998), 117.

140   **Some regions of China:** "Overview of Land Desertification Issues and Activities in the People's Republic of China," Food and Agriculture Organization of the United Nations, www.fao.org/docrep/W7539E/w7539e03.htm.

Chapter Five. Let Them Eat Caviar

146   **Thomas Jefferson's letter to James Madison:** Thomas Jefferson to James Madison, September 6, 1789, *The Founder's Constitution*, web edition, eds. Philip B. Kurland and Ralph Lerner (Chicago: University of Chicago Press and the Liberty Fund, 2000).

149   **Garrett Hardin's essay:** Garrett Hardin, "The Tragedy of the Commons," *Science*, December 13, 1968.

150   **first real data points:** National Science Foundation, press release 12- 070, "Ocean Acidification Linked With Larval Oyster Failure in Hatcheries," April 11, 2012.

152   **Korean entrepreneur Han Sang-hun:** Choe Sang-Hun, "Catering to Caviar Tastes From an Unexpected Place," *New York Times*, May 11, 2012.

155   **One U.S. study notes:** "Tim Pawlenty Says Every Child Is Born with a $30,000 Share of the U.S. Debt," *Tampa Bay Times*, Politifact.com, www.politifact.com

/truth-o-meter/statements/2011/apr/04/tim-pawlenty/tim-pawlenty-says
-every-child-born-30000-share-us-/.

169   **what we would call their nutrient manager, Recology:** IBM news release,
"Recology Teams with IBM in Quest to Help San Francisco Become First City in
North America to Achieve Zero Waste," May 31, 2012.

169   **The Van Gansewinkel Groep:** "Van Gansewinkel Groep," Van Gansewinkel
Groep, accessed November 6, 2012, www.vangansewinkelgroep.com.

172   **Kiehl's, the skin and hair care company:** "Recycle and Be Rewarded!"
Kiehl's, www.kiehls.com/Recycle-America/recycle-america,default,pg.html.

176   **The Ecological Sequestration Trust:** "Our Vision," The Ecological Sequestration Trust, video, 5:35, www.ecosequestrust.org/staticc/our_vision.html.

Chapter Six. The Butterfly Effect

183   **More than 900,000 schoolchildren:** Annenberg Foundation news release,
"Free App Connects Students, Naturalists Across the Continent as Birds and
Butterflies Go North," April 2012.

189   **organic food globally is now more than a $52 billion industry:** Data-
monitor, *Organic Food: Global Industry Guide 2009* (2008).

196   **70 million known organic and inorganic substances:** "CAS Registry
Number FAQs," American Chemical Society, www.cas.org/content/chemical
-substances/faqs.

206   **"Everything that is possible demands to exist":** Bertrand Russell, *A Critical
Exposition of the Philosophy of Leibniz: With an Appendix of Leading Passages*
(New York: Cosimo Books, 2008), 296.

207   **"What exists is possible":** Elise Boulding, "Toward a Culture of Peace in the
Twenty-first Century" (remarks given at the First Annual Global Citizen Award
Ceremony, Ikeda Center, Cambridge, MA, 2005).

Acknowledgments

Acknowledgments

We have been the beneficiaries of hundreds of generous and gracious people who have shared their creative gifts to inspire and inform the stories in this book. We especially delight in thanking all the people at our companies, our not-for-profit organizations, and our academic institutions, as well as our clients and collaborators, who have committed themselves to bringing these visions into reality for more than 30 years. They have given us a place to stand while leveraging beneficial change against the fulcrum of Cradle to Cradle thought.

We would like to honor our families for their loving support and patience as we have moved about the world doing our work. Bill would like to thank his wife, Michelle, and his children, Drew and Ava. Michael would like to thank his wife, Monika, and his children, Jonas, Nora, and Stella.

Our complete gratitude extends to the writers who helped us tell the *Upcycle* stories: Beth Rattner, Andy Postman, Lisa Williams, and Elizabeth (Biz) Mitchell. We thank Ken Alston, Christina Amacher, Michelle Amt, Kurt Andrews, Jay Bolus, Kira Gould, Tish Tablan, and Chris Wedding for their valuable contributions to the creation of this book. We also thank our editors, Duncan Bock at Melcher Media and Paul Elie and Sean McDonald at Farrar, Straus and Giroux, for their care and guidance.

And special thanks to Charles Melcher for his continuous leadership in improving the quality of publishing.

—William McDonough and Michael Braungart

About the Authors

William McDonough

William McDonough is a recognized leader in sustainable growth and development. Trained as an architect, McDonough works at scales from the global to the molecular through his advising, consulting, and architecture firms: McDonough Advisors, McDonough Braungart Design Chemistry (MBDC), and William McDonough + Partners. *Time* has called him a "Hero for the Planet," stating that "his utopianism is grounded in a unified philosophy that—in demonstrable and practical ways—is changing the design of the world." McDonough and chemist Michael Braungart coauthored *Cradle to Cradle: Remaking the Way We Make Things*, acknowledged as a seminal text of the sustainability movement, in 2002.

Michael Braungart

Dr. Michael Braungart is the founder and scientific CEO of EPEA Internationale Umweltforschung GmbH, an international environmental research and consulting institute headquartered in Hamburg. He is also the cofounder and scientific director of MBDC in Charlottesville, Virginia. He currently holds the chair for Cradle to Cradle® & Eco-Efficacy at Leuphana University of Lüneburg. He lectures at Erasmus University Rotterdam and holds the academic chair for Cradle to Cradle® Innovation & Quality at the Rotterdam School of Management (RSM). Dr. Braungart is also a professor and the chair for Cradle to Cradle® Design at the University of Twente and a visiting professor of building technology at TU Delft in the Netherlands. He has taught at Carnegie Mellon University, the Darden School of Business, the Bauhaus-University, and the University of Wales.

A Note on the Making of This Book

Upcycling concepts have guided the creation of this book. The materials used represent years of research and experimentation by MBDC, EPEA, and Melcher Media. Down to the molecular level, *The Upcycle* is as close as is currently possible to being fully optimized as a biological nutrient—a thing designed not only to do no harm but to be reintroduced into the environment in a beneficial manner.

The book was printed by Quad/Graphics in Leominster, Massachusetts, on 50# B18 Glatfelter Offset Antique finish cream-white paper that is Sustainable Forestry Initiative (SFI) Chain of Custody certified. Manufactured by Glatfelter in Spring Grove, Pennsylvania, using about 50 percent biomass for energy, this virgin paper is made with no added optical brighteners or elemental chlorine bleach.

The special cover material, Bio-Paper, was produced by C-Stone in Southern California and is made from a blend of limestone and plant-based resin. The paper uses a base of NatureWorks' Ingeo™ biopolymer, made from plants and not petroleum, which is Cradle to Cradle Certified™ Silver. Manufacturing Ingeo™ produces 60 percent less greenhouse gases and uses 50 percent less nonrenewable energy than traditional polymers like PET and polystyrene and contains no ingredients of concern to human or environmental health.

The inks used were assessed by EPEA and designed in collaboration with Flint Group in Europe with the goal of optimizing ingredients to be biological nutrients. The bookbinding adhesive, assessed by MBDC and produced in the United States, does not have any problematic intentional inputs.

The body type for this book is set in Tiempos Text, designed by Kris Sowersby for Klim Type Foundry. The callout type and cover type are set in Sharp Sans, designed by Lucas Sharp for Pagan & Sharp.